The **N**

SCIENCE

'Publishers have created lists of short books that discuss the questions that your average [electoral] candidate will only ever touch if armed with a slogan and a soundbite. Together [such books] hint at a resurgence of the grand educational tradition... Closest to the hot headline issues are *The No-Nonsense Guides*. These target those topics that a large army of voters care about, but that politicos evade. Arguments, figures and documents combine to prove that good journalism is far too important to be left to (most) journalists.'

Boyd Tonkin,
The Independent,
London

About the author
Jerome Ravetz is author of the classic study *Scientific Knowledge and its Social Problems* (1971, 1996). Formerly Reader in History and Philosophy of Science at Leeds, he also helped found the Council for Science and Society and is a pioneer in the area of science and safety. He is now an independent scholar and self-employed consultant working mainly on the problems of management of uncertainty in risks and environmental issues. His selected essays were published as *The Merger of Knowledge with Power* (1991) and with Silvio Funtowicz he co-authored *Uncertainty and Quality in Science for Policy* (1990). He is currently a Visiting Fellow at the James Martin Institute for Science and Civilization at the University of Oxford.

Acknowledgements
Thanks to my longstanding colleagues Silvio Funtowicz and Zia Sardar.

Other titles in the series
The No-Nonsense Guide to Globalization
The No-Nonsense Guide to Fair Trade
The No-Nonsense Guide to Climate Change
The No-Nonsense Guide to World History
The No-Nonsense Guide to Conflict and Peace
The No-Nonsense Guide to Human Rights
The No-Nonsense Guide to Animal Rights

About the New Internationalist
The **New Internationalist** is an independent not-for-profit publishing co-operative. Our mission is to report on issues of global justice. We publish informative current affairs and popular reference titles, complemented by world food, photography and gift books as well as calendars, diaries, maps and posters – all with a global justice world view.

If you like this *No-Nonsense Guide* you'll also love the **New Internationalist** magazine. Each month it takes a different subject such as *Trade Justice*, *Nuclear Power* or *Iraq*, exploring and explaining the issues in a concise way; the magazine is full of photos, charts and graphs as well as music, film and book reviews, country profiles, interviews and news.

To find out more about the **New Internationalist**, visit our website at
www.newint.org

The **NO-NONSENSE GUIDE** to
SCIENCE

Jerome Ravetz

The No-Nonsense Guide to Science
First published in the UK by
New Internationalist™ Publications Ltd
Oxford OX4 1BW, UK
www.newint.org
New Internationalist is a registered trade mark.

Cover image: Andrew Brookes/Corbis

Series editor: Troth Wells
Design by New Internationalist Publications Ltd.

 Printed on recycled paper by T J Press International, Cornwall, UK
who hold environmental accreditation ISO 14001.

British Library Cataloguing-in-Publication Data.
A catalogue record for this book is available from the British Library.

Library of Congress Cataloguing-in-Publication Data.
A catalogue for this book is available from the Library of Congress.

ISBN 10: 1-904456-46-4
ISBN 13: 978-1904456-469

Foreword

MY PERCEPTION OF science at school was similar to that of most other students. Theories were considered to be unchanging and offer exact predictions. I had to learn by rote many sets of facts. Like the students that Edmondson and Novak interviewed much later in their 1993 study of American colleges*, I certainly had a 'positivist' approach to science: first there's an idea, then a theory and then it becomes truth. Experiments were exercises that had a right answer which I failed to deliver most of the time.

Two things changed this. First, the experience of working in the laboratory of Dorothy Hodgkin on the structure of insulin in the 1960s, a project which she had started 30 years before. This was a real culture change. There were lots of debates about what we should do and what would work; nothing seemed obvious or certain.

But second, and more important, I became involved while working with Dorothy in local politics. As Chairman of the Planning Committee of Oxford City Council, I was required to make decisions when there were very few data and often what were available were uncertain and contradictory. I did not go into politics, but instead retired to the safe world of academic science. There I discovered that the art of 'successful science' was to identify problems that could be solved, and to ignore the others.

But that approach was soon challenged. My work took me to the world of medicines, to drug discovery. I suggested ways that putative drug targets might be inhibited. This was very exciting but it was only the start of drug development. Many drug candidates were abandoned because they showed signs of toxicity or even because they resembled other toxic molecules. In the face of an increasingly litigious society, drug candidates were withdrawn in clinical trials with the slightest danger signs. However, the uncertainties remained and many safe drugs proved to be dangerous in the long term.

A similar scene of uncertainty presented itself when I became involved in agricultural research policy in the early 1990s. Ministers assured us that their policy with respect to BSE (bovine spongiform encephalopathy) was based on 'sound science'. I worried that we knew very little about the prion or hypothetical infectious particle, that we could not measure it in most body fluids and that we knew nothing very much about how it led to the spongiform encephalopathies.

More recently my experience as Chair of the Royal Commission on Environmental Pollution has reminded me that ignorance characterizes most environmental policy. What does happen to molecules from sprays, paints, pesticides, fire protectants and thousands of other useful chemicals when they are released from products into the environment? What will be the effect of new supersonic aircraft on climate change? What does happen to other trophic levels of the food chain and to the wider ecosystem when we fish out one species?

This book begins to address such questions. It does so in a direct and often provocative way. It challenges our hopes for economic progress from genomics, robotics, artificial intelligence, neuroscience and nanotechnology in the coming years. It suggests that science may very often be characterized by malevolence and muddle. But it also recognizes that science can and will contribute to our health and wealth in the coming years. This is certainly essential reading for all aspiring scientists – and for some older ones too!

Professor Sir Tom Blundell FRS,
Department of Biochemistry,
University of Cambridge, England.

* Edmondson, K and Novak, J (1993) Cornell University study of American Colleges J Res SciTeaching. 30, 547-559.

CONTENTS

Foreword by Tom Blundell 5

Introduction 8

1 Science now..................................... 11

2 How science changed reality 20

3 Second thoughts from history.................... 34

4 Little Science, Big Science, Mega Science 47

5 Scientific objectivity 61

6 Uncertainty..................................... 78

7 Science and democracy 94

8 People's science 112

9 Science, its future and you 127

Contacts.. 133

Bibliography 135

Index 137

Introduction

FOR A LONG time science had been seen as the way to the real salvation of humanity; now it is also recognized as a possible instrument of our destruction. Making sense of this contradiction is the most important challenge for thinking people in our time. For this we will need to get beyond our inherited assumptions about science, its special sort of knowledge, and its interactions with society.

If we ask, 'what is science?' we discover a paradox. For 'science' is assumed to be objective knowledge, less dependent on personal opinion than any other sort of knowing. But the common answers to the question about science are enormously various, and they depend on the personal situation of the person involved.

For students, science is a huge pile of facts and techniques, requiring uncritical assimilation and mastery. For the general public, science offers the promise of curing disease. For the audiences of the media, science is a collection of intriguing or amusing ideas and gadgets. In earlier times, science could be a means of liberation from religious dogma or popular superstition.

For some, it now has negative aspects. It can be a means of gaining high profits or increasing the powers of the state. Animal experimentation can be condemned as insensitive or cruel. Tampering with life and reproduction violates our deepest sense of identity and sanctity.

Various science-based technologies are criticized as dangerous – nuclear energy or GM crops, for example – or as instruments of oppression, in the case of bio-prospecting or bio-piracy. And our total science-based industrial civilization shows strong signs of fouling its nest in the biosphere, with catastrophic consequences that might already be too late to avert.

Any reader will have their own reactions to the

elements of this list of opinions. None of these partial views of science is simply right or wrong. If we are to bring science successfully into the next phase of its history, we will need to cope with this diversity, and to comprehend its meaning. This book is designed to help with the development of that understanding. It looks at science as it is now, and shows how we got here. It explains how science can be genuine objective knowledge, and at the same time be conditioned by uncertainty and values. It also shows how our knowledge is selected and shaped by processes that reflect those values and the commitments behind them. And it highlights ignorance, which paradoxically is a crucial element in our understanding of the role of science.

In general, it promotes a 'post-normal' approach to science, which takes account of the uncertainty and value-loading that is now so pervasive. Further, it scrutinizes science from the point of view of modern democratic society. And finally it provides a list of questions for the reader, to help them create their own vision of science.

Recognizing the great achievements of science of the past and present, and the continuing great excitement and promise of the scientific endeavor, this *No-Nonsense Guide* aims to provide the elements of a new perspective so that the promise of science can continue to be fulfilled.

Jerome Ravetz
Oxford

Glossary of some of the terms used in this book

Genomics – the study and manipulation of life at the genetic level. A 'gene' is the unit of heredity; it is usually a component of the cell nucleus, acting with other genes and the intra-cellular environment.

Robotics – the technology that deals with the design, construction, operation and application of robots.

Artificial intelligence – the field of study that designs computers that can simulate or surpass human intelligent behavior.

Neuroscience – the science of the nervous system, especially the brain. We can now 'see' brain activity in real-time. The powers over consciousness that are being created raise the most far-reaching ethical issues.

Nanotechnology – technology that operates on a scale of a few nanometers, that is, a few billionths of a meter. It is potentially applicable to all fields of physical and life technology. But there are also very serious potential hazards of many sorts.

GRAINN – an acronym of all the above.

SHEE – the sciences of safety, health and environment, plus ethics.

M&M – malevolence and muddle, or the potential for misuse of science and for making mistakes within it.

Post-normal science – the sort of inquiry, usually issue-driven, where the facts are uncertain, values are in dispute, stakes are high and decisions are urgent. Its core ideas include an 'extended peer community' and the recognition of a plurality of legitimate perspectives on every issue.

The Precautionary Principle – advocates measures to anticipate, prevent or minimize adverse effects of scientific progress where there are threats of serious or irreversible damage. Lack of full scientific certainty should not be used as a reason for postponing such measures. ∎

1 Science now

Farewell to the old classifications, such as physics, chemistry, biology. Welcome to new ones, like GRAINN – short for genomics, robotics, artificial intelligence, neuroscience and nanotechnology.

THE FIRST THING we must do is to put the old subject headings in their place. We still think in terms of physics, chemistry and biology as ways of understanding the world in the relevant aspects. But to understand where the action is right now, and where the problems of the future are unfolding, we should identify the main focal points of development.

These areas are not at all limited by disciplinary boundaries. The distinctions between science, technology and public policy are vanishing. The new areas of science are conceived in terms of possibilities and problems. They structure the debates in which we all need to engage.

There is a convenient acronym for the state of play now, 'GRAINN', modifying GRAIN coined by Bill Joy of Sun Microsystems in the US. This stands for genomics, robotics, artificial intelligence, neuroscience and (eventually dominating it all) nanotechnology.

Science has always had multiple motives, including knowledge for its own sake, power over nature, and profit for developers. Now the main driving force for GRAINN is the hope for profits for industry – but the weapons developers are also looking for a harvest here.

We are already familiar with the promise of genomics. Knowledge of the genome opens up rapidly increasing possibilities for interventions and manipulations. We can identify genetic propensities to certain diseases and other sorts of vulnerabilities. Extending 'genomics' to all the interventions at the level of the cell and below, we have the new bio-engineering. All species, including ourselves, are subject to manipulation and modification. There seems no limit to what we can do in the way of altering nature and ourselves.

With robotics, technology has gone far beyond

simple production-line devices and mere playthings. Backbreaking human toil has become technologically obsolete – in the rich countries at least – thanks to machines. Soon much of present, boring, repetitive human labor – physical or intellectual – will be unnecessary and uneconomic. Furthermore, the possibility of robots that reproduce themselves is now on the agenda.

Once the limitations of human labor are bypassed, the future technologies of manipulation of energy and materials can scarcely be imagined. But what will happen to the majority of people who will have lost the status that is called a 'job'? For that is the only thing that now entitles anyone to a decent position in society. Here we have an impending social problem that cannot be ignored forever by the prophets of the robotics age.

Behind all this lies artificial intelligence (AI). After some decades of hype and frustration, the field is finally coming of age. No longer trying to simulate or replace human intelligence, these techniques focus on the sorts of things that digital computers – with rapid but linear processes – can do better than our brains, with slower but holistic processes. AI becomes the integrating technology behind both genomics and robotics. Also, we are on the brink of creating real 'cyborgs', where our bodies and minds are increasingly integrated with artificial devices.

And now coming in from science-fiction is nanotechnology, operating on the molecular, even atomic level. There is no branch of science or technology that cannot be affected, or even transformed, by the ability to work at that scale. With molecular-level 'assemblers', it has been claimed that we might soon have nearly free solar power, augmentations of the human immune system to protect us from or even cure the common cold and cancer, and bugs to clean up every sort of pollution – the list is endless.

Advances in instrumentation enable Neuroscience to advance at incredible speed. In real-time we can 'see' how the brain responds to stimuli of many sorts, even to our beliefs about its operation! The powers over consciousness that are being created raise the most far-reaching ethical issues.

If all this had been happening 50 years ago, there would be no containing the optimism and euphoria. 'Tomorrow's World' is coming today! But now the public, and even some of the scientists, are wiser, more cautious, perhaps even cynical, about these happy predictions. It is not merely that some great technological ventures, like civil nuclear power, have failed. Worse, there is a recognition that some will succeed in their own terms and then become not a blessing but a curse.

We might sum up these concerns with another acronym: M&M. The first of these stands for 'malevolence'. It is all too easy to speak of 'we' in connection with the benefits or the direction of science and technology. But there is very little democratic control over this great engine of knowledge and power. Giant corporations and giant state bureaucracies have their own agendas; and what is right for Monsanto or the Pentagon is not necessarily right for the rest of us.

People in these corporations and establishments may well have the very best of intentions for economic progress or for national defense. Yet their organizations can be ruthless or destructive and, in effect, malevolent.

There are other sorts of malevolence in technology as well. At the moment these are best known through the term 'hackers' who attack computer systems. So far, terrorists have been content with relatively unsophisticated methods; but if some terrorists became hackers, or vice versa, they would change the shape of global conflict and the prospects for civil society irreversibly. Anyone who designs or operates a technological system in ignorance of malevolence, is living in a bygone age.

The other 'M' stands for 'muddle', or perhaps for 'Murphy' as in 'Murphy's law' which states that whatever can go wrong, will. Among the prophets of technological change – even those who are pessimistic or frightened – there still seems to be an assumption that all the new systems will work as intended. This flies in the face of experience. Many of the current debates over technology focus on 'unintended consequences', some of which could not even have been anticipated when the technology first came into use.

As to computer software, on which all new technologies depend critically, it is known that most large software systems are completed at excessive cost and with lengthy delays, and fail to deliver the promised performance. Indeed many large software systems are simply abandoned. Internet pathogens ('malware') are scarcely under control, and there is no certainty that we can keep them at bay indefinitely. Muddle in design leads to vulnerability to malevolence by attackers, with ever increasing costs and hazards.

It is quite possible that some of the GRAINN technologies will be implemented on information technology systems that are overly complex, error-prone and vulnerable, and then released on people and the environment with inadequate testing and control. The possibilities of things going seriously wrong in a nasty way, due to malevolence or muddle, should not be neglected. The 'brave new world' of the GRAINN technologies will still be rooted in the messy, imperfect, conflicted and corrupted world we have inherited from the past.

Partly from an awareness of the M&M syndrome, new sorts of science have been developing. For them there is a very convenient acronym: SHEE (see glossary p 10). This has some delicious irony built in to it. For the issues of safety, health and environment are even now considered to be 'soft'; they don't make money for the firm or bring glory to the scientists. And as to ethics that isn't even Science at all! There is no Nobel Prize for Safety. SHEE science is women's work, rather like childcare and cooking. So we have the dialectical progression: in response to the challenges and threats of the powerful sciences now becoming led by GRAINN, we become acutely aware of the M&M syndrome, and in response we cultivate the SHEE sciences.

The challenges, and the responses, are not waiting for the full development of the GRAINN sciences.

To a significant extent, they are already there. The serious threat of global climate change can be seen as a case of M&M. Certainly, no-one intended that fossil fuels would produce a 'greenhouse effect' in the atmosphere.

That is a classic case of 'unintended consequences' or Murphy. For some years after the problem was first announced, there was a lively debate among the scientists. The skeptics – always a minority – were appreciated for their usefulness in finding weak points in the arguments. As evidence of climate change from diverse sources accumulated, the remaining 'skepticism' became 'denial'.

However, the fossil-fuel lobby continued its campaign unabated. At this point the issue changes from 'muddle' to 'malevolence', as those vested interests are only pretending to engage in a real scientific debate. In response to this new class of challenges, many leading scientists and scientific institutions change their orientation – on this sort of issue – from 'normal science' to SHEE, warning of the dangers and invoking lifestyle, politics and ethics in their arguments.

Along with climate change we have a host of similar issues. These range from obesity and diabetes resulting from junk-food addictions, to gender-bending pollutants that have a deadly effect in spite of their very low concentrations in the environment. Or there are the new sorts of microbial diseases, the results of globalization in its invasion of pristine tropical habitats and its production of mass intercontinental travel. Some of these problems involve science more directly, others less. But they are all characteristic of our modern science-based society. And it is clear that old-fashioned 'normal science' does not suffice for their solution.

The response to this challenge is already underway. The most serious of warnings have been given by Sir

Martin Rees, the Astronomer Royal. He is now Master of Trinity College Cambridge, one of the world's great centers of scientific excellence. In his book *The Final Century* he has shown how easy it would be for well-intentioned scientific research to be converted into a deadly threat to civilization.[1]

Another eminent figure, Sir Robert May, the President of the Royal Society of London, educates the scientific community in uncertainty and values, and assists in the development of democracy in science policy. The Royal Commission on Environmental Pollution, under Professor Sir Tom Blundell, has a distinguished record in highlighting the uncertainties and dangers in our situation. An Institute for Science and Civilization, reflecting the same concerns as this book, has recently been established at Oxford University.

With growing recognition of the threat, elements of a solution are being created. As soon as a potentially dangerous development is recognized, there is now a pressure group ready to warn against it. Dialogues are opened up, sometimes involving governments at the highest level. This new spirit of openness does not derive from pure altruism on the part of the promoters of new technologies. They have learned that if they lose the support of the public, the innovation may fail and their investment may be lost.

And the SHEE approach is becoming accepted in all sorts of unlikely places. After many generations of attempting to control floods by dams and straight channels, engineers in America and Europe have learned to work with rivers and not against them. They are restoring bends in the rivers and flood plains, and are redesigning cities so as to retain runoff water from rain rather than getting rid of it as quickly as possible. The philosopher Francis Bacon taught that 'Nature, to be commanded, must be obeyed', and this lesson is being remembered at last.[2]

Science now

Thus in many ways there is a growing new awareness in science. It is still a minority view in the broader community of science, technology and industry. It gains in support every time there is a painful event that demonstrates the inadequacy of the old simplistic understanding of science.

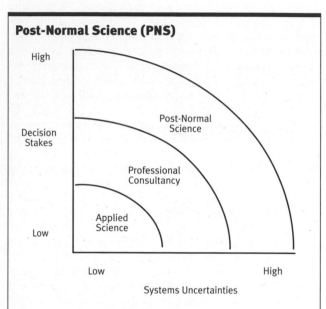

Post-Normal Science (PNS)

The inner zone corresponds to the routine, puzzle-solving 'normal science' described by the philosopher TS Kuhn in his book *The Structure of Scientific Revolutions*.[1] In that area, systematic uncertainties and value-loadings can be neglected in ordinary practice.

In the middle zone the professionals, such as engineers or surgeons, cope with uncertainties in their fields of practice, and with high costs of error or failure.

In the outer zone, of post-normal science, facts are uncertain, values in dispute, stakes high and decisions urgent.

All the policy issues raised by the GRAINN sciences are of this sort.

1 Thomas S Kuhn, *The Structure of Scientific Revolutions* (University of Chicago, 1962).

Within it are some crucial growth points of science as it evolves. It is not the whole of science; many exciting discoveries and beneficial applications are still being made in mainstream science. These are well described elsewhere.

In this *Guide* we focus on the new issues and challenges. Changing course to an appropriate, sustainable science and technology will not be easy or simple. We need to understand how they are quite different in style from the existing mainstream sciences.

We can call them 'post-normal', for they relate to this coming age when all the old comfortable assumptions about science, its production and its use, are losing their force.

It is not a question of simply replacing mainstream science and technology, exemplified by GRAINN, by the protective sciences of SHEE. It is a question of the relationship between the two styles, which is now out of balance to a dangerous degree. This is the great debate in which humankind is now beginning to engage and on which its fate can depend.

To talk about science as if GRAINN were all that mattered, and M&M and SHEE could be safely neglected, would be foolish indeed. With these new ideas we can articulate a new vision of what science could be all about. It would certainly be a more humane enterprise, in which awareness of uncertainty, and an appreciation of democracy, were cultivated.

1 Sir Martin Rees, *The Final Century* (London, Heinemann 2003) **2** Francis Bacon, *The New Organon*, Book 1, (1620), Aphorism 3.

2 How science changed reality

The origins of the contemporary Western idea of science can be traced to a few key epochs and events. For example, there was Darwin's idea about 'evolution' or Einstein's 'theory of relativity'. More recently we discovered 'the environment'. Today we face 'scientific perplexities'.

HUMAN BEINGS ARE different from all other species in the way that we not only change the world around us, but also change ourselves rapidly in the process. We can trace the evolution of humanity over hundreds of thousands, even millions, of years, thanks to our discovery of material remains of past cultures. Our knowledge of what people were like in those prehistoric times is of course very scanty; we must make inferences from the scattered, accidental remains of their material productions.

But when we come to historic times, within the last 5,000 years or so, some records of civilization and culture are available. We can actually make reasonable guesses as to what people believed, thought and even felt. Within that period, we can trace major developments in human understanding. Not all of these come from science; but many do. Since ours is now a science-based civilization, the contribution of science to our reality is of crucial importance for our understanding of ourselves.

The 'Greek miracle'
Although technical mastery over Nature had been increasing steadily through the millennia, there is something distinctive about Science as we now understand it. And that distinctiveness was born with the Greek-speaking peoples, living on the Greek mainland, on the islands of the Eastern Mediterranean, and on what is now the Turkish coast around the 6th and 5th centuries BCE.

It seems that around then, some great thinkers began to imagine knowledge as something worthwhile for its own sake, starting to become independent from ordinary practice on the one hand, and from religion and magic on the other.

There is a great legendary figure, Pythagoras, who perfectly expresses this new movement. In addition to the pursuit of learning, he was also a political reformer and the founder of a mystical brotherhood. For him, there was probably no separation between the three activities. But he gave us a new conception of mathematics, as the study of number, shape and structure in themselves. Before his time there had been some very sophisticated practical mathematics, including quite refined calculating methods and an appreciation of general rules for problem-solving. And the use of numbers for magic, which even now has not quite vanished, was then just commonsense.

Pythagoras, however, looked at numbers in a more general way, proving their properties independently of the size of any particular number. Later, people did the same with 'geometry', transforming it from 'earth-measuring' to become a general study of shape and structure, where actual size is irrelevant.

Pythagoras also applied mathematics to the world of practice; he discovered that the pitch of a musical sound is related to the size of the instrument. We see this most easily by 'stopping' a vibrating string halfway along, and getting the simplest harmony, the 'octave'.

After Pythagoras, all branches of science developed among the Greeks. The greatest scientist of them all, Aristotle, investigated many fields in the biological and social sciences, as well as philosophy. Moreover, he actually created our idea of an organized, disciplined inquiry, rather than just debating issues endlessly, as did his teacher Plato. After Aristotle, the demarcations between science, magic and religion were blurred again. There were some very great scientists in later antiquity, but the creative genius of classical Greece was not renewed.

The Copernican Revolution

A proper history of science would include an account of the crucial contribution of the Islamic civilization to the evolution of science, along with an appreciation of the special excellence – and great contributions – of the characteristic sciences of India and China.

But our concern here is with the formative epochs when advances in science changed the reality that people in 'the West' now experience. With that acknowledged Eurocentric bias, and remembering that the common sense in the future might be different again, we move on to Europe in the early modern period, known as Renaissance and Reformation.

Until very recently in human history, it got really dark at night, both outdoors and indoors. The starry heavens were then an intimate part of ordinary experience. Their various changes were observed and recorded; and they were an obvious topic of scientific inquiry, usually tied in with religious belief.

Thus, both the Jewish Passover and the Christian Easter are defined in terms of the full moon and the first day of Spring. They fall on different days in the two religions because each has its own method for calculating the predicted occurrence of those events. But all the computational methods, based on the nearly-cyclical behavior of the heavenly bodies, eventually fall out of step with the actual motions. The correction of time-measurement is a perennial problem for science, one that is still with us.

For all those earlier millennia, people assumed that the heavens were stretched out over the earth, which itself was flat. In Greek times, scientists established that the earth is spherical, and hence the heavens as well. But it was still common sense to imagine that the heavy earth was at the center, and the illuminated heavens rotated around it. This fitted in with religious ideas, particularly the Christian view. In that, Hell is the hot place down below – as we know from volcanoes – and Heaven is the outermost, unchanging place where God and the angels reside.

In the 16th century the world for Europeans expanded suddenly. New worlds were discovered in 'America', and someone actually sailed around the globe. And a Polish astronomer, Nicolaus Copernicus, published a book that announced two very implausible ideas.

One is that the earth is spinning like a top. The other is that it is also flying through the heavens. At first this all seemed quite ridiculous. The 'sound science' of the day, as well as theology and common sense, were all against it. But Copernicus's treatise was a heavyweight of technical astronomy, which promised to provide accurate predictions at last.

By his time, there was a long-standing scandal over Easter. This was defined in terms of the sun and the moon, being the first Sunday after the first full moon after the 'vernal equinox' – when the day becomes as long as the night. There were computational methods whereby Easter could be calculated, on the basis of the predicted motions of the sun and the moon, for any year into the future. But these predictions had become very inaccurate and so anyone could see that the real 'first Sunday after... etc' was not the same as what the computations prescribed.

Also, no astronomical tables got the positions of the planets right, and so 'mathematical physic', or astrological medicine, could not make progress. On the basis of its promise for the improvement of both religion and science, Copernicus's theory was taken seriously, in spite of being so bizarre.

By the time that Galileo Galilei arrived on the scene, the whole idea of that hierarchical Christian universe was beginning to fade. Hearing about the invention of a 'spyglass', in 1609 he designed and constructed a telescope that enabled him to see things never seen before. These included mountains on the moon, the phases of Venus, the moons of Jupiter, and thousands of previously invisible stars. The *Starry Messenger,* where he announced these discoveries, is one of the most thrilling scientific stories of all time.[1]

These amazing discoveries made him absolutely convinced that Copernicus was right. But convincing others was not so easy. Many years later, he published a great *Dialogue* on cosmology.[2] It was a literary masterpiece, but a scientific disaster and a political catastrophe. Hauled before the Inquisition, the aged Galileo recanted. He was humiliated, but the Catholic Church had trapped itself in opposition to a theory that very soon became a recognized scientific truth. The Scientific Revolution was built around Copernicus's ideas. With its success the counter-intuitive motions of the earth became an obvious fact.

The completion of the Copernican revolution was left to others, like Johannes Kepler and Isaac Newton. They showed that scientific astronomy needed the assumption of the earth's motions. And common sense was changing fast. University textbooks and popular science books increasingly accepted Copernicus's theory as obvious. By the end of the 17th century, the earth-centrists were as marginalized as the flat-earthers and the astrologers. It had taken nearly two centuries for the Copernican revolution to be completed, but the real world was – at least for educated Europeans – now very big and very empty.

Paradoxically, the direct proof of the Copernican theory was slow in coming. In the early 18th century the English astronomer James Bradley made observations of the stars that indicated that the earth, moving in orbit around the sun, came closer to them at one time of the year and then further away a half-year later.

And only in the mid-19th century did Jean Foucault's pendulum enable everyone to imagine themselves seeing the earth rotating beneath their feet. These delayed 'confirmations' are a reminder that scientific knowledge is a complex system, involving many more aspects than simple data and simple theories.

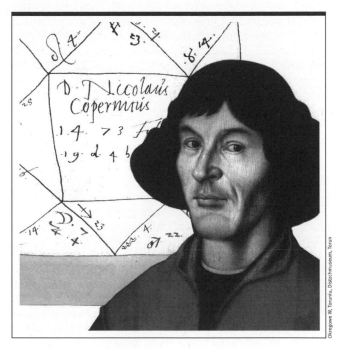

Okregowe W, Toruniu, Distzchmuseum, Torun

The Scientific Revolution

This is the epoch that made science as we now under-stand it. But it was not the result of a single discovery; rather it came from the interaction of the work of some prophetic visionaries, with some key achievements of science, and with a changing educated common sense. It was later said that modern science was a result of the precepts of Bacon, the concepts of Descartes, and the experiences of Galileo.

Francis Bacon was a lawyer, philosopher and would-be savior of mankind. His criticisms of science were mainly based on its corrupted state in his time. Hardly anyone cared about producing real knowledge, and those who did were largely immersed in their own fantasies. He hoped to get royal patronage for an institution where the sincere and disciplined scientists would do their work. This required playing the courtier's game, a dangerous enterprise in which he failed. But after his death his works became an inspiration to scientific reformers for centuries. His combination of personal morality with an 'inductive' method of careful observation and unbiased analysis appealed to those – like the British – who distrusted vain speculations.

René Descartes was a very great mathematician and philosopher. His disillusion with his humanistic education led him to seize on mathematics as the key to truth. His achievements there were immortal; they included the perfection of algebra and the invention of coordinate geometry. And his systematic philosophy of the world changed things forever. For him everything was reduced to matter-and-motion; and on that basis he hoped to construct physics, biology, medicine, psychology and ethics. The wonder is that he got as far as he did.

Galileo was the most practical of the three, aside from his obsession with proving the Copernican theory. He created modern physical science. He understood as no-one else the procedures – and some of the pitfalls – of relating the conceptual objects of a mathematical theory with the material things of experiment and craft practice. His work on mechanics, including the analysis of uniformly accelerated motion, cleared away many confusions and enabled Newton to construct his complete dynamical system.

What these three very different people had in

common was a commitment to a very special sort of knowledge. They rejected the enchanted world of the magicians, while retaining their dream of power. And they rejected the bookishness of the university teachers of the time, while retaining their appreciation of the need for disciplined learning.

There was never a single doctrine of 'the Scientific Revolution'; the innovators and their successors each went their own way. Their work interacted with the rapidly changing, educated common sense of Europe. There needed to be a rather 'disenchanted' world view among the public for it to be accepted at all; and then its key texts helped the process along.

Charles Darwin

After the Scientific Revolution, it seemed that God was being pushed back into a corner, being needed only on Sundays. But many scientists still believed that a 'great Designer' was needed for the creation of all those intricate and exquisitely designed organisms, human and of all other species. It just didn't seem plausible that such elegant structures could have arisen by chance combination. Yet amidst all this evidence of some superhuman intelligence, there is the equally strong evidence of pain and evil rampant in the world. How were these two perspectives to be reconciled?

Charles Darwin was influenced by such great philosophical issues, but his scientific focus was on the narrower question of the way that new species come into being. It was obvious from the geological record that they did, and it was really implausible to imagine that God had personally created so many millions of different sorts of life.

But there was no direct evidence, as no-one had ever seen a new species of life emerge. Darwin had to rely on different sorts of indirect evidence, none of which was conclusive. The geological record showed no 'missing links' between different life forms. Artificial

selection operates on the weaker, domesticated varieties of animals, in that humans may decide which
animals to breed and which to keep. Nevertheless,
Darwin achieved the great insight that in some way
Nature also 'selects', by favoring those individuals
that are better adapted to survive to adulthood and
then to breed.

He labored long and hard, and finally published
his ideas in 1859. He had to rush at the end, when he
discovered that a colleague – Alfred Russel Wallace
– was getting there first. Then came his big book, *On
the Origin of Species*, that made a sensation among
the reading public. There was much scientific criticism, which Darwin handled for about a decade. Then
he just withdrew from the fray.

Among the working scientists, the idea of evolution
by natural selection didn't seem to help very much
until the early 20th century. Then it was combined
with the nearly-forgotten genetics studies of Gregor
Mendel and the foundations of contemporary scientific genetics were laid.

Later in Victorian times, the biological theories
of Darwinism were widely confused with the social
philosophy of Herbert Spencer – 'the survival of the
fittest' – and it was also used as an apology for unbri-
dled individualism. Also, his later work on *The Descent
of Man* served to discredit the Adam and Eve story of
the Bible. It was this attack on some Christians' religious beliefs, more than natural selection in general,
that later provoked the religious reaction called 'fundamentalism'. This tendency then promoted 'Creationism',
presented as an alternative theory to Evolution, which
still strongly influences education in the US.

It is correct to say that Darwin discovered a profound
and simple scientific truth, which profoundly changed
our common sense picture of our place in nature and
our relation to the deity. But the history of Darwin's
theory of Evolution by Natural Selection is anything

but simple; and it is unwise as well as inaccurate to pretend that it is.

Einstein and the Pandora's Box

Coming to the end of the Victorian age, there was a great sense of satisfaction and optimism. 'Progress' was the dominant theme. Hand-in-hand, science and technology had brought vast improvements in all spheres of life. With our hindsight, we know that Europe was then on course for a devastating war; and strange things were already happening in science.

In the year 1905, there was a failed revolution in Russia, a rehearsal for the one that would soon destroy an empire. In the same year, a patent office clerk in Switzerland sent three papers to leading physics journals. One explained the 'Brownian motion', an apparently perpetual random motion of very small particles. This was widely accepted as conclusive evidence for the existence of 'atoms', and it settled a debate that had raged among chemists for generations.

Another dealt with a very specialized problem of the sorts of energy emitted by a hot 'black body'. This was important for theoretical thermodynamics; someone had already fitted the experimental data by making an odd assumption about energy being transmitted in packets and not continuously. Einstein developed this systematically, and thereby laid the foundations for 'quantum theory'. Finally, in a very abstruse study of 'the electrodynamics of moving bodies', he came up with a radical revision of physical reality. The only constant in his scheme of things is the velocity of light. All motions of bodies are 'relative', in that it is impossible to find an absolute framework against which they can be measured. The mass of a body increases as its velocity approaches that of light. And an odd equation appears in the mathematical analysis: $e = mc^2$. This indicates that there is some sort of equivalence between matter and energy.

How science changed reality

It took quite a while for Einstein's work to be appreciated. 'Relativity' in a more general form became a sensation just after the end of World War One. Astronomers confirmed Einstein's prediction that the light from a distant star would be bent by the sun's gravity. And as scientists wrestled with the paradoxes of atomic structure, a strange sort of 'uncertainty' was revealed as fundamental to our knowledge. Atomic structure, based on quantum ideas, was eventually married to Einstein's relativity theory through the fateful equation between mass and energy. Less than half a century after Einstein's three papers, the atomic bomb was born at Alamogordo in New Mexico. The world has not slept easily ever since. The image of the mad scientist whose arrogance might destroy the world is no longer a curiosity of literature. It has become a central element of popular culture.

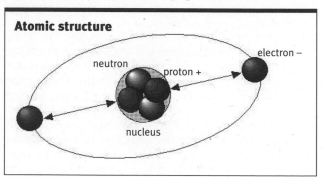

Atomic structure

neutron · proton + · electron − · nucleus

The Environment

All sorts of great advances have been made in mainstream science since the end of World War Two. But a real change in our perception of reality has come through a variety of sources at the fringe. For generations there had been prophets warning that our industrial civilization was unsustainable in one way or another, either destroying farmland and forest, or exhausting key resources, or polluting ourselves and our environment. They went unheard. In the early postwar period, the euphoria about science extended to the environment. Just as atomic energy would make electricity too cheap to meter, the new pesticides (starting with DDT) would, many scientists believed, kill all the bad bugs and preserve all the good ones.

Then, as an inauguration to the traumatic 1960s, was published *Silent Spring* by Rachel Carson.[3] This told a simple tale of a sudden absence of songbirds; and it explained this through the effects of pesticides in destroying their habitat and poisoning their young. In America, the reaction was instantaneous. The industry interests tried to refute the message and also to destroy the messenger; but they failed. And idealistic young people picked up on 'the environment' along with civil rights and peace, as one of those things symbolizing what was wrong with the consumerist culture they were simultaneously rejecting and enjoying.

There were also some pictures that became truly iconic. One was of the great tail of a whale just starting its dive. How could we continue to butcher such magnificent friends, to no purpose at all? The other was a totally unintended by-product of the Space Race: a picture of Spaceship Earth, delicately shrouded by its cloud cover. In an age when the medium was rapidly becoming the message, such images, reproduced as posters on millions of young persons' walls, defined the new consciousness. Also, they put the environment on the moral high ground. This caused much chagrin among those Americans who sincerely believed that 'When you've seen one redwood tree you've seen them all', and equally sincerely asked: 'What's posterity done for me?'.

The characteristic sciences of the environment involve strong commitments to the highest human values, and they recognize the deep uncertainties in our situation. In that way, they are radically different from the 'normal' sciences where uncertainty is tamed and values are suppressed. With them we see post-normal science in action.

Scientific Perplexities

The advent of the GRAINN sciences (genomics, robotics, artificial intelligence, neuroscience, nanotechnology) as conditioned by M&M (malevolence and muddle) has made the problems of the environment seem simple and almost reassuring in retrospect.

In many cases we can identify what is wrong – perhaps a particular set of pollutants – and then create the elements of a practical solution. The barriers are frequently no more than simple greed and stupidity, which only take time and perseverance to overcome.

The first wave of environmental problems, arising out of Rachel Carson's *Silent Spring*, responded quite well to this approach. With further progress came more sophisticated problems, such as pernicious chemicals in trace concentrations, or destructive practices – like unsustainable logging – that are promoted by corrupt governments in poor and rich countries alike. More recently we have come to see how much our lifestyle keeps us locked into bad practices; such as relying on fossil-fuels for universal private transport, or using a large proportion of our primary food production for feeding farmed animals and fish.

But now the problems become ever more complex and tangled. For an example, we can consider the use of human embryonic stem cells in research. Some find this unethical in itself, while others object because it involves the use and then the destruction of the embryos.

Science will not be able to define the point at which a developing embryo suddenly becomes 'human'. Yet the emergence of 'human-ness' is a subject of great significance for ethics and policy. Even moral theology is accepted as a valid perspective in that debate. Further, there are those who accept the research as ethical, but who are afraid that it will enable the cloning of humans.

Of course, that practice is illegal in many advanced countries. How could it be made illegal everywhere? Even if it is, there is a huge international trade in all sorts of things, for which being illegal is only a minor inconvenience. Could there be effective regulations and sanctions to prevent the cloning of people from becoming established?

Even if the established scientific community takes fright at the prospect of illegal cloning of humans, how could a ban on research and development be enforced? Journals could refuse to publish papers in the field. But there could be a 'shadow' research sector, whose scientists are driven by some combination of pride and greed, and possessing its own means of communication. Would there need to be an international scientific police force to prevent the abuse of this technology? How could it be made immune to intimidation and corruption?

If we have good reason to fear the science-fiction scenarios of black markets in gene-designed babies, or of armies of robot-like identical clones serving dictators, then we may need to stop the development of the basic science immediately, on these purely pragmatic grounds. This particular form of cloning may not be the most urgent policy problem confronting us now or in the near future. But it shows how the consideration of a typical nascent GRAINN technology can involve a complex of issues ranging from theology to the Mafia. Whatever might be the solutions to such problems, they will not be determined by the science alone, and they will not be simple or easy. That's why we can call them 'scientific perplexities' and with them, we are well inside the post-normal age.

1 Galilei Galilei, 'Starry Messenger' (1610) in *Discoveries and Opinions of Galileo*, translated by Stillman Drake (1957) 2 Galileo Galilei, *Dialogue on the Two Great World Systems*, 1633, translated by Stillman Drake (University of California, 1953). 3 Rachel Carson, *Silent Spring* (1962; Houghton Mifflin 1993)

3 Second thoughts from history

What about the errors of science? We find these in the work of Pythagoras, Galileo, Newton, Lavoisier and even Darwin. How much more interesting science teaching would be if students were shown this human side of science.

SINCE WE ARE concerned with the future of science, we should remember that the present is just a moment between the past and the future. And when we are open to the possibility of the future being really different from the present, we can learn many things from the past. The standard history of science, which I sketched in the previous chapter, is one of unremitting, perhaps even unrelenting progress, nearly up to the present. This account was quite plausible a century ago, but it is high time that we had another look at it.

Other ways of doing science

We might start back towards the beginning, whatever that might be. Through all the millennia of prehistory, there was nothing special about Europe. If any region deserves credit for being the birthplace of humanity, it seems likely to be Africa. Archaeologists have traced the development of implements, from rough chipped stones to refined tools. They have also analyzed the gigantic stone monuments which, all over the world, seemed to serve both astronomy and religion.

About a century ago, people discovered caves in Southern France, first at Lascaux, with the most amazing paintings of men and animals; and they were firmly dated to be more than 20,000 years old. At first the scientific experts, the archaeologists, were sure that they were fakes: it was impossible that primitive cave-dwellers could have had such refined artistic sensibilities. But when they finally were accepted as genuine, other problems arose. For as the pictures were interpreted, it became clear that for those people, there was no distinction between science, art, religion and magic.

Generations of propagandists for modern science had taught – on the basis of their own experience – that the shackles of religious delusion had to be cast off, in order that science could progress. Yet here were people who were sharing and transmitting their techniques of animal-hunting through group rituals of combined worship and hunting practice; and in the course of that they were producing high-quality art.

Coming into historical times, we have various great monuments whose construction we still cannot understand. Most famous of these is the Great Pyramid of Egypt. The erection of this structure is astonishing in itself, but when we reflect that it is not merely a pile, but was built around complex tunnels and chambers, we are left bewildered. What sort of architecture was required, to make such shapes within shapes? There is no trace of earlier, somewhat simpler structures on which the skills were learned.

On another front, the early agriculturalists of the Western Hemisphere gave us a rich legacy of edible plants, some of which, like maize, are quite remote from their wild ancestors. These were not the result of chance; why shouldn't we call their creators 'scientists'?

More troubling for the triumphalist view of European science, we are now becoming aware of other scientific traditions in which other aspects of reality have been cultivated to a high degree of excellence. Take acupuncture and yoga; are they just superstition and placebo? Or perhaps they belong to alternative systems of scientific knowledge which are quite effective in their own way for healing, in spite of deriving from totally different philosophical backgrounds.

It appears that in those other systems, some things were even better understood than in our scientific culture. The Sanskrit writings of ancient India show a refined awareness of consciousness and its varieties, that far exceeds ours. Among the ancient Egyptians, and also the Tibetan Buddhists, there are doctrines of the psychological processes of dying, compared to which our own understanding is primitive in the extreme.

Through all the centuries of European conquest, all such alternatives were despised and suppressed. In many colonies, traditional medicine was made illegal by the Western imperial masters. Now we are beginning to discover, not merely our debt to other civilizations, but how much we can still learn from them.

But civilizations come and go. For some centuries Western Europe north of the Pyrenees was backward and barbarian, while the Mediterranean lands of the Islamic civilization flourished. Our debt to their science and technology is enshrined in words like 'muslin', 'alcohol', 'chemistry' or 'algebra'. As we have now come to understand it, classical Islamic science had its own intellectual and ethical framework, derived from the moral injunctions of the Koran. Zia Sardar has written extensively on this.[1] Further east, in their own ways the Indian and Chinese civilizations produced great and progressive science, which we have only recently come to appreciate through therapies like yoga and acupuncture.

Typical of the debt to the East, and of Europe's failure to acknowledge it, is the list of the 'Three Great Inventions' with which Francis Bacon explained the rise of European power during the Renaissance. Little did he know that printing, gunpowder and the magnetic compass all derived from the East, with transmission and perfection under Islam. One of the great issues of world history is how these three imported inventions became the tools of the European conquest of the world.

In modern times, we found ourselves on a rising exponential curve of discoveries and inventions. And

so it seemed just a century ago, when Europeans and their science dominated the world. It was not just about sheer power; progress meant prosperity and freedom too.

But then came a couple of world wars, the atomic bomb and environmental degradation. Now we are wiser, and beginning to be humbler. We are becoming ready to learn. The great lesson of history is that there is not just one way to do science. It need not be the exclusive property of experts, and it need not be restricted to what can be cut up and counted. The future is open, and is ours to make.

Some errors from which we can learn a lot

During the last half century we have learned that the applications of science can be misdirected or harmful. But we are still shielded from the awareness that on occasion science itself can be mistaken. The idea that competent and well intentioned scientists can get it wrong and then persist in their errors, is still subversive. For this, our inherited science education is largely to blame. For those who are subjected to it, it amounts to a most insidious form of brainwashing. For year after year, students never see a mistake, except for their own. The infallibility claimed by the Pope is nothing compared with the implied infallibility of the textbooks.

Even for those who go on to research and learn by painful experience that some facts are less factual than others, this remains as a sort of secret dirty truth. It is as if teachers are good shepherds for the students, protecting them from the perils of doubt. Indeed, we can say that science education is the last remaining unchallenged dogmatic teaching in the modern world. How long can it last? And when discovery, disillusion and reform come to science, how radical will they need to be?

What a world of excitement, insight and creativity is lost, by our collective inability to confront error in real science! The materials for a study of creative error are available. They have been provided by the historians of science even if not exactly trumpeted abroad by them. Along with the errors of commission in science, there are the errors of omission, where the relevant expert community rejects a result that is later shown to be correct. If we see how great scientists can err, that opens a window to the creative process. And if we admit how scientific communities can err, that could induce a welcome humility about the scientific consensus at any time.

The errors go back to the beginnings of mathematics with Pythagoras. We recall that he believed that Number is the key to everything. That would include shape, in particular the symmetrical structures, such as triangles and polygons, which even now still have some magical significance. If Number rules all, then the ratios of their various dimensions should be capable of being expressed by whole numbers, as 1, 2, 3 and so on.

But disaster struck; there is a legend that one of the Pythagorean brotherhood discovered a ratio that could not be expressed in whole numbers. It may have been the square root of two ($\sqrt{2}$), relating the hypotenuse (long side) to a short side of an equilateral right triangle. Or it may have been the ratio of the diameter to the side of a pentagon, the shape at the center of a five-pointed star, and a very powerful mystical symbol. Either way, that first mathematical vision was shattered. Ever since, quantities like $\sqrt{2}$ have been called 'irrational' – crazy.

Pythagoras made another fruitful error in his theory of musical harmony. He created a 'rational' musical scale based on repeats and inversions of the basic ratios 2:1 and 3:2, the octave and the major-fifth. This universal harmony was intended to replace the multiplicity of 'modes', the musical scales which each separate city-state used for its religious ceremonies. Unfortunately, the tones on Pythagoras's scale do not work out perfectly. Different methods give different results; the error has been known through history as 'Pythagoras's comma'. Also, his scale includes some intervals that are neither convenient to play nor pleasant to hear.

These imperfections were important many centuries later in the history of science. Galileo's father was a musicologist who argued against the Pythagorean numerological approach to harmony, and some of Galileo's earliest experiments were on music. His own aesthetic work – including theories of musical harmony and of architecture – strives for a new relationship between mathematics and the empirical world.

The very richest source of scientific errors is to be found in Galileo himself. His great, far-ranging genius touched on so many themes that speculation and error could not be segregated off from the established facts. Even the immortal discoveries of the moon were a strange mixture of insight and error. The drawing of the moon (see below) shows both.

At the top we see the points of light which Galileo correctly interpreted as mountains. The sketch shows an isolated peak on far left of the crater. About three-quarters of the way down, there is a large, perfectly round crater, whose far side almost looks like an amphitheatre. Galileo made special mention of it, comparing it to Bohemia, now a part of the Czech Republic. Unfortunately, no-one ever saw it again.

Galileo's moon

The flat spots Galileo interpreted as seas, the soft edge of the sphere as atmosphere, giving rise to theories of 'men on the moon'.

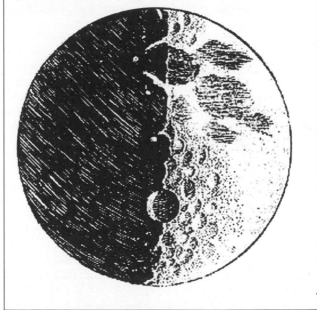

Galileo interpreted the flat spots on the moon as seas. Also, he explained the apparent smoothness of the moon's surface near the edge, as the possible result of a layer of atmosphere. Together these theories implied the possibility of a life-support system on the moon. This was dangerously close to the theory of men on the moon that was advanced by various Protestants. This was a known theological debating point, for such men could not have heard the gospel of Christ, and hence the Roman Catholic Church was not universal. Only a decade previously, a notorious adherent of the Copernican theory, Giordano Bruno, had been burned at the stake for numerous heresies. Galileo's enemies watched him closely.

Galileo had other troubles with his telescope. At the magnifications he used, his lenses were at the very limit of their reliability. Many of his lenses – manufactured by himself, by a secret method – were just not good enough, and his opponents could point to spurious images to discredit the whole story. And he had a problem with the stars; they did not seem any larger or closer when seen through the telescope. This was seized on by his opponents: why should the moon seem larger when the stars didn't? His explanation involved inaccuracies of vision inside the eye.[2]

Galileo's great work, *The Dialogue on the Two Great World Systems*, was an attempt to disprove the traditional system of the world and establish that of Copernicus. To establish the two motions of the earth, he used the model of a round bowl of water being swung around. When that is done, a hump of water swishes around the bowl. This he interpreted as the tides on the earth. Unfortunately, there are two tides a day and not one. Also, he simply did not have a theory of mechanics whereby the motions could really be explained. So he relied on his wonderful command of language to poke fun at the Aristotelians and argue for the plausibility of his ideas. In that great book he showed how difficult it can be to distinguish the genius from the crank, in a man whose ideas are ahead of their time.

Galileo's greatest mistake of all was the way he quoted the methodological principle of his longstanding protector and friend, Pope Urban VIII, to the effect that God can arrange things however he likes. For this came at the very end of the book, as if to sum up the whole argument – but it was put into the mouth of the Aristotelian scholar 'Simplicio'. If he had planned to make a fool of the Pope, he could not have done it more effectively. He lost the Pope's protection, and his enemies then had a free hand.

Galileo's great English contemporary, William Harvey, became embroiled in debate over his theory of the circulation of the blood. On the one hand, he had what seemed to be conclusive quantitative evidence, based on the volume of blood that is expelled from the heart every minute. But the path of the circulation was not entirely clear. In particular, critics queried how blood got from the very small arteries to the very small veins. Exasperated, Harvey insisted that he had seen and probed large open connections, 'anastamoses', between the two systems.[3] No-one else ever found them; and we now know that the pressure difference is enough to drive the blood through the capillaries.

We also find that Newton's genius did not protect him from making the wrong guess on an important question of physical theory. He wrestled with the paradoxical properties of light. On the one hand, the way it travels in straight lines indicates that light is 'corpuscular'. But he also knew of phenomena that indicate a periodic property. After all, 'Newton's rings' are a standard experiment for demonstrating that light travels in waves. The colored bands we see on CD disks are also evidence for waves as they are explained by 'interference'. Newton did worry about the alternatives, and finally came down for the one that was rejected in the 19th century when advances in mathematics enabled the proof of the wave theory of light.

But the progress of science always brings surprises. In the early 20th century, experiments indicated that light has *both* corpuscular and wave-like properties. Breaking with the picture of the world inherited from Descartes, Nils Bohr invoked the Chinese philosophy of complementarity, or *yin-yang*, to make sense of these paradoxical phenomena.

A really interesting error is enshrined at the foundations of modern chemistry. When Lavoisier created a rational system of chemical nomenclature, he named one of the basic elements 'oxy-gene', meaning the acid-former. He did this in spite of the fact that this is also the agent of combustion, and so might well have been called 'pyro-gene'. (One of his co-discoverers, Scheele, actually called it 'fire-air'). But Lavoisier was convinced that all acids are created by combustion, and therefore consist of an 'oxidized' base combined with hydrogen. He had prepared phosphoric acid, sulfuric acid and others, in this way.

Unfortunately for Lavoisier's theory, not long afterwards Humphrey Davy showed that the base for 'muriatic acid' is not a product of combustion, but is a true element, which he called 'chlorine', as it is a green gas. Ever since, students of chemistry have imbibed Lavoisier's error with the name of Oxygen (in German it is Sauerstoff, the sour-stuff); but of course no-one is so tactless as to point it out to them.[4]

Even Darwin's theory of evolution by natural selection had a bumpy ride. Once we began to know about the fossil record, it became clear that enormous stretches of time would be required for the very gradual processes of natural selection to create man out of the primeval slime. And Lord Kelvin, a distinguished physicist, used the accepted theory of the cooling of the earth to show that not very much time was available. Given the measurements that indicated the heat of the earth's iron core and the rate of cooling, a mere 40 million years, or even less, had elapsed since the earth was molten.[5] For such a broad and general theory as Darwin's, this was as near to a refutation as one could get. Darwin was certainly worried by it.

Nearly half a century after the publication of Darwin's *The Origin of Species*, another great physicist, Rutherford, showed that radioactivity – recently discovered – was keeping the earth warm. Kelvin's calculations were inaccurate. Belatedly, it was established that there was time for Natural Selection after all.

An overly strict interpretation of Darwinism had unintended beneficial consequences for literature. The mystery of lichens, those tiny plants that grow on rocks, had been explained quite soon after Darwin's book, in terms of a co-operation between algae and fungi. But this lapse from Darwinian individualism was treated with scorn and derision by the botanical establishment. At the end of the century, the theory was reinforced by a set of detailed drawings of the microscopic structure of lichens. The author, an Englishwoman, was a brilliant naturalist, with very good connections in the scientific community. But through a combination of dogmatism and misogyny, she was denied a hearing. So Beatrix Potter withdrew from science, and created Peter Rabbit. It took a few decades for symbiosis to become scientifically respectable; but Beatrix Potter never returned to what she called the 'grown-up world' of science.[6]

Some thoroughly justifiable scientific errors were exposed by Einstein's work of 1905. When it was understood in Victorian times that electricity, magnetism and hence light travel as waves, the task was to obtain the properties of the medium that is undulating. Researchers produced ever more refined models of the 'luminiferous aether', and its properties became ever more paradoxical. With Einstein's special theory of relativity it was established that there is no such medium anyway. Similarly, the perturbations in the orbit of the planet Mercury were explained by the gravitational pull of another planet, named Vulcanium, which is so close to the sun as to be invisible. Again, the calculations of its position never worked; and Einstein's general theory of relativity gave quite another cause for the phenomenon.

The most notorious recent scientific error of omission concerned the theory of Alfred Wegener, that the main continents had once been joined. The outlines of the continental shelves of Africa and South America fit so well that they might have been bits of a jigsaw puzzle. There is other geological and paleontological evidence as well. But Wegener's evidence was dismissed by scientists with the sort of contempt that is usually reserved for parapsychological experiments.[7] Only long after his death was 'plate tectonics' discovered, and now it is obvious scientific common-sense.

The point of all these examples is most definitely *not* to show that scientists can be stupid or prejudiced. Rather, it is to serve as a reminder that any scientific belief may turn out to be incorrect in some important respect. Scientists, like everyone else, must use tests of initial plausibility to filter out weird ideas. Sometimes these tests can mislead, as with the motions of the earth or of the continents. If the very greatest scientists can err, there is no need to claim infallibility for the rest. Nor should the discovery of error be a cause for blame or shame.

Of course, no-one likes to be proven wrong; but to admit error is the beginning of true understanding. Perhaps one way to bridge the gap between the sciences and the humanities, and to make science teaching interesting again, would be to involve students in reviewing instructive errors of the great scientists, along with the impassioned debates on the great scientific issues of their time.

The Quakers have a fundamental principle: never forget that you might be wrong. How much better the world could be, if this principle were built into science education. Then science might become an instrument of nonviolence, of the sort that the world so desperately needs.

1 Zia Sardar and Merryl Wyn Davies, *The No-Nonsense Guide to Islam* (New Internationalist/Verso, London, 2004). **2** More detail on Galileo's problems is found in my earlier book, *Scientific Knowledge and its Social Problems*, (1971, 1996). **3** Steven A Lubitz, MD, 'Early Reactions to Harvey's Circulation Theory: The Impact on Medicine', in *The Mount Sinai Journal of Medicine*, Vol. 71, No. 4, 2004; http://www.mssm.edu/msjournal/71/71 **4** References to the impact of Davy's discovery are not common. One account is: Ian Hacking, 'Science Turned Upside Down', in the *New York Review of Books*, Vol 33, No 3, February 27, 1986; A sequel is: http://www.nybooks.com/articles/5001. **5** On Kelvin and the minimum age of the earth, see a standard account: http://geowords. com/histbooknetscape/ko6.htm. The pro-Darwinian bias is obvious. **6** Tom Wakeford, *Liaisons of Life* (Wiley, 2001). **7** http://www.ucmp.berkeley.edu/ history/wegener.html

4 Little Science, Big Science, Mega Science

The social organization of science is now very different from the way it was in the era of 'little science' a century ago. 'Big science' came in with the Bomb and has now been supplanted by 'mega science' – creating a new set of problems.

THERE'S ANOTHER ILLUMINATING way to consider the history of science. This is to study, not so much what was discovered as the way in which the work was done. This is particularly important now, for the issues of the governance of science are crucial. What sorts of research gets done, and what doesn't get done, who decides on priorities, and what happens to discoveries once they are made, are the big questions concerning science of our time.

The issues that are expressed by the acronyms GRAINN, M&M and SHEE (see **At a glance**) are the pressing ones for us. This new situation has come on us so quickly that the common understanding of science as a social activity is still lagging behind. History can help our understanding, in showing how the present has been emerging from a more familiar past.

At a glance

GRAINN – genomics, robotics, artificial intelligence, neuroscience and nanotechnology

M&M – malevolence and muddle

SHEE – safety, health and environment, plus ethics

PNS – Post-normal science: the stage we are at today, where all the old comfortable assumptions about science, its production and its use, are in question ∎

Little Science, Big Science, Mega Science

Little science

This term hearkens back to a period, roughly extending up to World War Two, when science was generally seen as a small-scale enterprise, led by independent persons and largely self-governing in its priorities and in its procedures for quality-assurance. There were of course many changes during that period and many varieties of practice, but in Europe at any rate this was the image of science. If anyone were to ask 'Who runs the activity of science?' the natural answer would have been: 'The scientists, of course. Who else?'.

The scientists were not only independent; they had a certain nobility about them. They needed it; a lifetime in science was generally a vocation rather than a career or a job. Few posts were available and there was little promise of fortune, power or even fame.

Another sure sign of the high moral tone of the scientific community was the way it managed its quality-assurance. Whatever the philosophers might have said about the infallible Scientific Method, real scientists know that 'discovery' is hard and uncertain work. Further, it has always been far too easy to convince oneself that a result is sound when in fact it is vitiated by errors. So it is necessary to have assessors to 'grade' scientific reports, to see whether they are worthy of publication.

In other fields of activity the examiners come from outside. In fine art, there are professional critics who are accepted as better judges of quality than the community of artists themselves. But in mainstream research science that system of external review could not work. For in research science, each project is embedded deeply in a context of other results involving abstruse technicalities. To identify the hidden errors and to assess the promise of a new piece of work requires someone already familiar with the field.

So science has always used 'peer review' by colleagues for its quality control; the scientists assess each other's work. It's not difficult to see how readily this could be corrupted if scientists widely adopted a policy of 'I'll scratch your back if you scratch mine', or if powerful people used their influence to get favorable evaluations.

In spite of all the inevitable cases of error and corruption, that system of colleague peer review has worked remarkably well down through the generations. And that could happen only because the community of scientists has upheld the system and because in the last resort they have believed in their work. Paradoxically, it turns out that the work of discovering the objective facts about the natural world has depended quite critically on the motivation, morale and morality of those doing the work, even more strongly than in most other fields of human endeavor.

So 'little science' was characterized by idealism – and also by isolation. Although some branches of science were closer to practice – such as chemistry and the fields related to medicine or to the management of natural resources – the overall impression was of scientists freely offering their discoveries to each other and to the world. If someone in a practical area could make good use of them, so much the better; but that was not what science was about.

This self-understanding of science is reflected in various ways even now. Thus, textbooks in all but the most directly applied fields do not include 'context' in any but the most limited fashion, perhaps a page or two at the beginning and at the end. Also, science writers still tend to make a distinction between the good news of scientific discovery – which it is their job to celebrate – and the bad news of environmental problems, which are left to a lesser breed of gloomy 'environmental journalists'. And the publicity campaigns to promote interest in science have been based on the principle 'to know us is to love us', as if the lone scientist is still in there, holding up a test-tube to the light to see if it has the chemical that will cure cancer. Even if much research is still an individual endeavor, motivated by the highest ideals, the social context of science has been transformed in ways that make the old concepts of science obsolete.

Big science
Whatever the realities on the ground, there is no doubt that the self-made image of science went through a profound and partly traumatic transformation during World War Two, particularly in the Manhattan Project in which the first nuclear bomb was created. Atomic physics had been one of the most aristocratic of fields and it had operated on resources that in retrospect seem almost comically small.

In the later 1930s Enrico Fermi did some fundamental experiments in a little fishpond on the terrace outside the window of his lab. It provided the water that he needed for protecting the reaction. Not long afterwards, he was involved in a project on a massive industrial scale, whose outcome was fateful for all humanity. With the bomb, 'big science' was born.

To their credit, many scientists were deeply concerned at what they had wrought, and afterwards devoted much effort to the attempt to contain its effects. Some found it difficult to comprehend how the search for truth had actually produced something capable of such evil. But many of them felt a sense of personal responsibility, even guilt, and a variety of movements were created. Foremost among them was Pugwash, named after the birthplace of the industrialist who financed it. Pugwash succeeded in maintaining a dialogue between Western and Soviet scientists through all the worst times of the Cold War.

By the 1960s, other issues arose that further eroded the idea of the essential innocence of science. One was 'the environment'. It had, after all, been scientists who had enthusiastically invented and promoted the pesticides that caused the 'silent Spring' made famous by Rachel Carson's book of that title. Another was the war in Vietnam, where the American public realized that some nasty, even possibly criminal, weapons had been developed by scientists – including academics – for the US Military.

An iconic case of social responsibility was that of a leading biochemist Arthur Galston, who as a graduate student before the war had discovered a chemical that promoted growth in certain plants. He was later annoyed to learn that someone had made a commercial weed killer out of it, and was drawing profits from the patent. But then he was horrified to discover that the military were using this chemical to defoliate large areas of Vietnam and in the process were poisoning many innocent civilians. He did the honorable thing and joined the campaign against Agent Orange.

The campaigns for Social Responsibility in Science date from the later 1960s, in response to this new and troubling awareness. They accomplished much, but in a sense they were already too late. For in the postwar period science became big.

It was much bigger in overall size, projects became larger in scale, and it was much more closely linked to the military and to industry. The governance of science altered subtly but significantly. When the cost of a project became so much bigger than the salary of a scientist plus an assistant or two, support could no longer be dispensed informally, as through university funds. Instead, external bureaucracies, usually state agencies with specified missions – or, for medicine, large private charities – were needed. Inevitably, scientists tended to respond to these external priorities rather than to their own, although of course there was a lot of room for maneuver.

Even on the lab floor there were important changes. With large budgets to manage, large numbers of staff of varying grades, expensive equipment to manage, and a never-ending search for new money, research units came to resemble industrial enterprises more than groups of scholars.

There was no place, except on the margins, for the lonely scientist with a vocation to discover his special sort of truth. Big, or industrialized, science was a career, one which could be very rewarding but which also had the drawbacks of being a manager in a demanding and insecure enterprise.

Mega science

The transition from big to 'mega' science came with two developments, one inside and the other outside the research enterprise. After 1945, the queen of the sciences was physics. After all, it was physicists who had made the bomb that won the war against Japan, and they would now create the Atoms for Peace program with its promise of everlasting cheap energy for all.

The American physicists also offered their wisdom to policy institutions like RAND, where mathematical simulations and digital computers would – they believed – solve the problems of American military strategy and global statecraft. All they asked in return was a series of ever more powerful, and ever more expensive, particle accelerators that, they also believed, would soon reveal the secrets of matter in its most fundamental states.

But after a couple of decades, physics was looking stale. The civil nuclear energy program never recovered from the accident at Three Mile Island in Pennsylvania. Furthermore, the particles displayed by the high-energy accelerators got ever more strange and enigmatic, while the cost of the accelerators was itself in a state of steady acceleration. And by this time biology was stealing the limelight – the new, hi-tech biology that exploited the discovery of the Double Helix of DNA.

But something else was happening to science. The postwar 'military-industrial-scientific complex', against which President Eisenhower had warned, was beginning to fade. In the Thatcher-Reagan years, private enterprise was taking over. Science was being harnessed to the profit motive. And in the fashion of that time, this was short-term profit, reflected in the quarterly balance sheet and the share price. The great industrial labs of the earlier 20th century, like the Bell labs where so many inventions were made, were disbanded.

Science was becoming just another factor of production. It was only the first, somewhat more speculative phase of R&D – or research and development. Of course, the process of privatization was not uniform; there were still some significant enterprises in the not-for-profit sector, noticeably the Human Genome Project. But even this endeavor had to cope with sharp competition from a private-sector organization.

Through all these changes, military science continued to prosper. In Britain it absorbs more than a third of the science budget but is hardly ever discussed as a policy issue. In the US, by contrast, it includes extravaganzas like 'Star Wars' and its descendants, in which it is difficult to know where the hardware ends and the fantasy begins.[1]

The overall size of the scientific enterprise continued to grow; and the scale of individual projects leapt in magnitude. As ever greater masses of data required ever more complicated calculations, the power of Information Technology soared away, while the costs plummeted. Computer simulations were developed, first to operate where experiments were impossible, then as a complement to experiments and eventually as a substitute for them. The lone scientist with her or his test-tube was, in retrospect, something from the horse-and-buggy age.

As mainstream mega science became ever more

Going for broke

Applications for patents have been rocketing. Under the World Intellectual Property Organization's (WIPO) Patent Co-operation Treaty companies can file a single patent application which would be valid in many countries. Each such patent has the power of scores of individual patents.

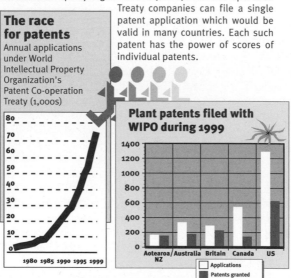

The race for patents

Annual applications under World Intellectual Property Organization's Patent Co-operation Treaty (1,000s)

80
70
60
50
40
30
20
10
0

1980 1985 1990 1995 1999

Plant patents filed with WIPO during 1999

1400
1200
1000
800
600
400
200
0

Aotearoa/NZ   Australia   Britain   Canada   US

☐ Applications
■ Patents granted

However, patents granted in individual countries are more numerous. And the US Patent and Trademark Office leads the world.

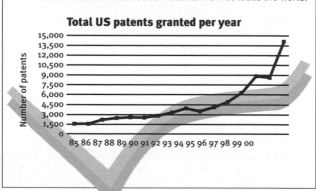

Total US patents granted per year

Number of patents

15,000
13,500
12,000
10,500
9,000
7,500
6,000
4,500
3,000
1,500
0

85 86 87 88 89 90 91 92 93 94 95 96 97 98 99 00

UNDP / NI 349

integrated with global big business, its ways of operating necessarily followed suit. For example, the intellectual property embodied in a scientific discovery could be free to all, only so long as no-one could make any serious money out of it. Now, in 'hot' fields the lawyers assess a result from the lab before the journal referees get a look at it.

Treating science as property is not new in corporate R&D. When research is tightly organized around particular well-defined company products and processes, such protective measures are inevitable. Industrial scientists have long been accustomed to preserving their professional integrity while operating within such constraints. But now the area of 'corporate know-how' is rapidly encroaching on fields of more exploratory, basic research, so that 'public knowledge' is ever more confined and constrained.

Even if someone wants to do 'little science' she or he cannot escape from the property system. Many products and processes are now captured by patents or patent applications, so that anyone is at serious risk of having their research interrupted or made prohibitively expensive. They themselves might be subject to lawsuits for patent infringement. As part of the globalization of the economy, the patent system – mainly in the US but also in Europe – has been transformed to favor those with the deepest pockets. Incredibly broad patents are granted on the most speculative of developments. The holders, with their platoons of expensive lawyers in reserve, then demand high fees from users and defy anyone to challenge their monopoly.

The global knowledge economy depends on a captive science in all sorts of ways. Ever seeking new forms of property, the multinational corporations send out scientists to discover possibly useful plants and animals in poor countries. Once their usefulness is established, the processes depending on them, or even the organisms themselves, are patented in the rich countries. The world's poor are then denied the opportunity to develop them their own way, or sometimes even lose the rights to their own plants and traditional processes. The Neem tree of India, known as the 'village pharmacy', has been a victim of scores of patents appropriating its use and products. This practice is made easier by a feature of American patent law, where 'prior use' abroad is invalid unless there is a publication defining it. Small wonder that this is called 'biopiracy'; and mega science is at its core.

There are severe difficulties in grafting the activity of mega science onto a practice that still has folkways from its 'little' phase. Mega science may dwarf old-fashioned little science in size; but it still needs it. As we have seen, quality assurance in research science depends on peer review and this work is very vulnerable to pressures of all sorts. It depends on commitment and morale; but how are these to be maintained under the new, commercialized conditions of mega science?

Recently, we have been encountering major problems with quality assurance in relation to examinations for students. The question 'who examines the examiners?' has become urgent as the system creaks and occasionally fails.

With mega science, research is no longer a private matter among specialists. The fate of big investments can ride on a favorable evaluation of research; and the temptations and pressures on evaluators increase all the time. We have recently witnessed the failure of quality assurance in other parts of the commercial world, as evidenced in the 1990s dot.com bubble swindles. Can we be confident that the institutions controlling mega science will play this game any more honestly?

Closely related to quality assurance is integrity. When those at the top of any social system lack integrity it will go corrupt; and science is no exception.

Cell lines fiasco

It is known that the traditional systems of quality assurance in science are under strain. Both the refereeing of papers and the peer-reviewing of grant applications place ever heavier burdens on already overworked scientists. Rivalries between research groups, along with commercial constraints and inducements, make the preservation of integrity ever more challenging. The whole subject is very difficult to investigate with social-science methods. So when there is a well documented story of a long-standing unsolved problem in a typical area of science, we can take this as evidence of a systemic difficulty in the quality assurance of science.

The area in question is defined by its materials: pure descendant lines of human cells in tissue-cultures. This biological material is an essential ingredient in research in a great variety of fields. The reliability of that research depends critically on the assured quality of the cell lines. One would think that systems of quality-assurance of materials, such as the ISO standards that are in force over all spheres of administration and industry, would be applied here with special rigor.

Astonishingly, this is not the case. Instead, we learn that for some 30 years scientists have known that human cell lines are very vulnerable to contamination by one particular strain. This is called HeLa, after the African-American woman Henrietta Lacks who died from a uniquely virulent cancer and whose body cells were then used for science. The scientist who discovered the widespread contamination, William Nelson-Rees, tried to alert his colleagues. But their reaction was negative; and his proposals for reform were never implemented.

At a recent conference in London it was estimated that a quarter of research using cell lines is worthless because of this contamination. And of course we don't know which quarter it is! In spite of this, there is as yet no sign that referees, journals or funding agencies wish to overhaul their procedures. It is not a happy scene. ∎

Andy Coghlan, 'Imposter cells are wrecking research', *New Scientist* 20 September 2003.

Integrity in mega science poses serious problems for the overall operation of the global knowledge economy that depends on it. For that new knowledge economy is constantly creating new dangers along with its benefits, most noticeably in the GRAINN fields discussed above.

The issues are essentially 'post-normal', as they involve uncertainty and ignorance, as well as issues of power, values and lifestyle. When science is deployed, the 'facts' alone are rarely conclusive. Their proper interpretation depends on the integrity of the scientists who debate them. But as scientists are increasingly absorbed into the mega science system – and become ever more dependent on industrial or government funding – their integrity is under strain. It is increasingly seen to be compromised. In many cases involving SHEE issues the public doesn't trust them when they offer reassurances of safety. In this way, mega science is destroying its most precious asset, the public trust in science that it needs if it is to have consent for its projects.

Mega science has tendencies that erode trust even in its ordinary working. In the days of little science most research could be done on modest means, some of it needing a patron. Then with big science, an investor of some sort – private, state or charitable – was needed. With mega science, much research needs venture capital to get started.

This is speculative investment, that demands a quick return. Scientists are under great pressure to make their results seem exciting, either to get the support or to keep it. The careful reporting of facts in specialized journals is being displaced by extravagant claims in the media. Of course, sometimes the claims are justified and great benefits are achieved. But when the public sees scientists behaving like PR men, this trust is modified accordingly.

Finally, for those contemplating working in mega science, you can forget about vocation, and even a career is now becoming problematic. Under present commercial pressures, research is increasingly at best a job, and frequently an ill-paid and insecure job at that.

Hype and 'megaphone science'

Professor Steven Rose is one of the very few scientists who combines eminence in research with fierce independence as a critic. He has described our condition as 'megaphone science' where there is no point in announcing anything less than a supposed 'major breakthrough'.

Genetics is especially susceptible to this form of corruption, since a story about the discovery of 'the gene for' some noteworthy condition is bound to attract interest and support. The most notorious case here is the 'gay gene' of 1993. Later we had the 'aggression gene', first identified in a Dutch family and then in some nasty lab mice.

Not all such genes are bad news; a group at Princeton found that genetic changes in certain special mice could alter a particular brain chemical associated with learning in mazes. Immediately, the prospects for super-children, on the model of Brave New World, were widely discussed. Steven Rose calls for the media to refuse to be overawed by men and women in white coats. Journalists should adopt critical, investigative journalism in the area of science, just as in all others. Only in that way can the scientists themselves be protected from the pressures for short-term profits by biotech and info tech companies. ■

'Time to look critically at science', *The Science Reporter*, July/August 2004.

Research is coming to be organized around temporary projects in which particular skills are mobilized. At the end of the project, the human resources, along with the material ones, are dispersed or disposed of.

Whenever two hi-tech firms merge, the labs of the weaker partner are no longer needed for copy-cat and patent-busting work. The labs are shut and most of the research workers are dismissed. Even in the universities, an increasing proportion of the researchers are temporary staff. They live from contract to contract, until their seniority makes them too expensive and they are discarded.

In former years, there was a steady stream of idealistic young people who went into science in order to make the discoveries that would benefit humanity. Nowadays, there may also be some idealistic young people demonstrating outside the lab, protesting against supposed cruelty to experimental animals, or against some aggressive practice of globalized mega science.

The clear and unchallenged moral superiority of the scientific vocation becomes a distant memory. For those who are concerned about the ethical aspects of a career in science, the organization Scientists for Global Responsibility offers excellent advice.[2]

Unfortunately, the image that is projected by the publicity for science is still about little science, perhaps with a nod to social and environmental concerns. It is badly out of touch with the dominant reality of mega science. Bringing the image up to date will be difficult and painful. Many good people have invested their talents in the science of an earlier epoch, and it will not be easy for them to admit that the reality has been pulled out from under them. For some who have already achieved scientific eminence, these new and troubling developments have become a challenge of the highest order to their leadership. With their independence and influence, they have a chance to make a real difference to the evolution of science as it passes into the next phase of its history.

1 See my essay 'Hardware and Fantasy in Military Technology', in JR Ravetz, *The Merger of Knowledge with Power / Essays in Critical Science* (London, Cassell/ Mansell, 1982), and on the website http://www.nusap.net.
2 See website: http://www.sgr.org.uk/ethics.html for the worldwide network: http://www.inesglobal.com. See **Contacts** for details of this and similar organizations.

5 Scientific objectivity

There is a strong tradition, going back to Galileo, that science is objective, impersonal and value-free. This is responsible for its strength and certainty. Post-normal science rocks this notion and suggests that instead of faith in 'objectivity' we should cultivate 'integrity'.

IN THE PRECEDING chapters we have seen that science has been constantly evolving throughout its history, and also that it has always been confronted by difficulties and challenges.

The recent transformation from little science through to mega science has happened with great rapidity. Many senior scientists are still in a sort of culture shock as their world has been transformed around them.

Through all this confusion, can we say that there has been some single change that is more important, more fundamental, than all the others? It is important to try to identify this essential change. By becoming aware of it, we can better manage it, and so guide science through the turbulent times ahead.

We have seen how the social organization of science has been transformed, in scale and in its external involvement. Has there been a corresponding recent transformation in the core practice of scientific inquiry? To identify this, we need to analyze the image of science that traditionally has been purveyed to science and the public alike – and then compare that to the realities that we now perceive.

To start, we can make a contrast between science and the study of history. History tells a story; it tries to capture the richness and contradictions of human life and striving. Of course, history uses disciplined methods of analysis of evidence, and it will even employ 'scientific' techniques like statistics where appropriate. But its end-product is a story, written in prose, about humanity.

We can appreciate the difference between history and science by considering the intermediate case, that of the social or 'behavioral' sciences, like economics or psychology. These are also concerned with people and the way they behave. But in general, these sciences operate by abstracting from real life. Their basic procedure is to consider general theories and then test them against standardized data. The variability and richness of ordinary experience, which may be crucial to an historical narrative, are excluded from their scope. Although there are many debates about methods within all those fields, the dominant tendency for a long time has been to promote 'objectivity', in imitation of the successful natural sciences.

There is a good reason why the 'softer' disciplines strive for objectivity. For it is almost universally agreed that natural science achieves a knowledge that is superior to that of arts or humanities disciplines. Unlike in arts subjects, a scientific statement is proved or disproved by methods that don't depend on human attitudes or cultural attributes. A scientific fact is the same in all places and cultures. Also, science is immune from the effects of human prejudice and of subjective values. Indeed, the methods of science are deliberately designed to neutralize the effects of the desires of the investigator. These two aspects of objectivity, impersonality and value-neutrality, are accepted as defining real science and hence genuine knowledge. They also contribute to its other basic quality, that of supposed absolute certainty.

This view of science is deeply entrenched. It can be found in a famous proclamation by Galileo himself in the *Dialogue on the Two Great World Systems*:

> *'If this point of which we dispute were some point of law, or other part of the studies called the humanities,*

wherein there is neither truth nor falsehood, we might give sufficient credit to the acuteness of wit, readiness of answers, or greater accomplishment of writers, and hope that he who is most proficient in these will make his reason more probable and plausible. But the conclusions of natural science are true and necessary, and 'l'arbitrio humano' has nothing to do with them.'[1]

I have left one crucial term in the original Italian. We now distinguish two senses of 'arbitrio', as in the very different words 'arbitrary' and 'arbitrate'. One refers to will and the other to judgment. For Galileo, both senses were present. He contrasts the presentational skills of the non-scientific disciplines, with the rigorous constraints of science. By his 'true and necessary' we can understand: objectivity. The human being, with his will and his judgment, is out of the picture.

Galileo's faith is reinforced by the practice of science teaching down through the generations. Traditionally, descriptions of laboratory work exclude the first person. Students learn to avoid, 'I poured the liquid' and to write, 'The liquid was poured'. That is supposed to be more 'objective'. The lesson is that the individual   afflicted by all sorts of imperfections – shouldn't count in the process.

Similarly, there is no place for judgment. Students have years of intensive training in solving standard preset puzzles. Each of the puzzles has just one correct solution amidst all the incorrect ones. The good, diligent student will find that one correct answer and receive a reward. There is no place for the exercise of judgment; indeed, that would only confuse the student and result in a lower mark.

In the ordinary practice of routine research, it is possible for a scientist to continue with this simplified picture of his work. There will be judgments at a technician's level which are required when equipment

Those troublesome 'outliers'

When we reflect on what actually happens in scientific inquiry, the illusion of total objectivity is quickly dispelled. Suppose that we are doing an experiment, obtaining the response – in some physiological indicator – to different doses of a toxic substance. Most of the data points cluster nicely, climbing up along a 'trend line' at a steady rate. This gives the 'dose-response relationship', that will help in the determination of the 'safe limit' for the substance.

But some points are 'outliers' – they do not conform. For example, there might be one point that shows a very small response to a very large dose. Is it only an artifact, some sort of error in the experimental technique or in the recording system? Or is it real? We must make a choice; there is no escape.

If we include it in the analysis, and calculate the 'best fit' trend line, the low outlier point drags the trend-line down. If it is actually unreal and we have accepted it, then we make an underestimation of the danger from the toxicant. Safety standards based on our interpretation of that data set would expose people to danger. On the other hand, suppose that the low outlier datum point is real and we ignore it. Then we will class the toxicant as more dangerous than it really is. The resulting safety standard will be too strict; it will thereby waste resources and – like all false alarms – can also lead to distrust of the system.

That's the dilemma. We must choose which risk to run when we choose a policy. There's no way of knowing whether the outlier datum point is real or not, without doing the experiment all over again and hoping that there are no outliers that time. And suppose we do, and there are some? We need to make a decision on what to do about the outliers, based on our personal experience of that sort of experiment. We may also need to consider the further consequences of the possible errors, in this case being either too lax or too strict in setting a safety standard.

'Objectivity' won't get us very far in that sort of reasoning; well founded judgments are necessary at every step. Galileo's great pronouncement doesn't apply here.

Of course, we know that in practice, setting a safety standard depends on the integration of the results of many such experiments. But that procedure will depend on assessments of the quality of the researchers, their institutions and the journals, all depending on personal experience and judgments. There may well be outliers occurring even in these aggregated data. In the practice of 'normal' science, a sort of 'objectivity' does eventually emerge from the synthesis of such judgments. But it comes from a fallible social process, and not from Nature in some unmediated way. The management of outliers shows how even in the ordinary practice of 'normal science' there is a touch of the post-normal. ■

does not work as it is supposed to do. But all the big issues have been settled for him or her, and the puzzle-solving researcher does not even need to know that they are there in the background. So complete is the illusion of objectivity, that even when judgments are quite essential in the ordinary practice of science, they are not recognized as such. A good case in point here is 'outlier' data.

The other side of 'objectivity', the supposedly value-free character of science, is indoctrinated more as a part of the implicit background to science education. It is just about inconceivable for a student to challenge the syllabus, asking why this particular set of topics has been chosen for a science course and not some others. Yet anyone who has been involved in a collaborative work of creating a syllabus knows of the fierce debates that are inevitable in that work. For there is never enough time to teach everything that is desirable or even important. There is a perennial clash of values, as each specialist fights for what she or he knows is essential for a proper course.

But for the students, the syllabus is just there, as if written in stone. Criticism of it is just inconceivable. The value-commitments that determine choice, first in science education and then in scientific research, are hidden from view. They are part of what the philosopher Thomas S Kuhn called the 'paradigm', the implicit unquestioned and unquestionable framework in which teaching and 'normal' research is conducted.

We are now in a position to assess the price that has been paid for the dominance of this sort of 'objective' science. It is not merely that the social sciences sometimes look like caricatures, pretending – or even trying to convince us – that there is no real difference between people and molecules. Worse, the reductionist approach produces a mindset where natural-scientists become uncomfortable with complex problems that cannot be reduced to research puzzles. These call for judgments to be made, and that is outside the rules of the game they have learned and played successfully. Such scientists feel happier when the really big, challenging problems of our age are carved up into little pieces for puzzle-solving research. But the complex realities of policy-related research cannot be reduced to the sums of such simplified parts.

The exclusion of real people from the scientific picture has its own severe costs. For the challenges of the environment and of sustainability involve people, with all their unpredictability and contrariness. The historian's insights are needed here as much as the scientist's expertise. But many scientists assume that only a fully trained expert has anything to contribute on the solution of problems with some scientific content, even when the problems are totally enmeshed in political and ethical issues. Those scientists who become involved in policy-related issues frequently have a learning experience that may be more deep and difficult than anything they experienced in their formal education.

The endeavor of Albert Einstein is the great example of the glory of 'objective' science and of its passing. For he drew great spiritual nourishment from the study of the impersonal, immutable laws of nature, so different from the squabbles of ordinary humanity. But out of his profound discoveries came the atomic bomb, which shattered the isolation and purity of science forever.

The decline of the illusion of objectivity

Over the last half-century, science has experienced great transformations in its scale, size, power, destructiveness, and corporate control and social responsibility. There is lively debate over many policy issues concerning safety, health and the environment, and over proposed innovations such as those in the

GRAINN set – that is, genomics, robotics, artificial intelligence and nanotechnology.

But until we get over the illusion of objectivity of science, as embodied in its supposed impersonality and value-neutrality, those debates will be hindered and distorted. So long as each side in a debate believes that it has all the simple and conclusive facts, it will demonize the other and dialogue will not be achieved.

We need not fall into some nihilistic philosophy of total subjectivity or power-games in science. That is not the only alternative to the lost illusion of perfect objectivity. To find a viable alternative we will need to examine why scientific objectivity is no longer common sense.

The process of growing beyond 'objectivity' is already well underway. At an accelerating pace through the later 20th century, we have become aware of crises and issues where the simple application of textbook knowledge has been radically insufficient. After the discovery of pollution, and the loss of credibility of civil nuclear power, we have increasingly encountered problems with a scientific statement which do not have a simple scientific solution.

Some of the SHEE problems are just obvious, but others are genuinely obscure. Just now there is controversy over mobile telephones and the radio masts that serve them. This radiation is very weak, and does not cause 'ionization' in the way that radio-active decay does. So there is a strong presumption that it is harmless. And if the effects are subtle and slow in being realized, there are huge difficulties in settling the question experimentally. Yet there are grounds for concern, since the human brain uses weak electrical signals which could possibly be affected by those being emitted by the phones or the transmitters. Galileo's faith in objectivity will not help us here. In such cases as these the 'conclusions of natural science' *for policy* are not 'true and necessary'. And there is no escape from the presence of 'human will' in the interests that are affected by the policy, nor any

---

Which energy system?

'Safe and cheap' was the sales pitch for nuclear power. Neither has proven to be the case.

Costs of electricity:
(cents per kilowatt hour with external costs)

Hydropower	2.40- 8.70
Wind	4.05- 6.25
Natural gas	4.40- 9.00
Coal lignite	6.30-19.80
Biomass (wood etc)	8.00-12.00
Nuclear	10.20-14.70
Solar PV	25.60-50.60

- Fossil fuels and nuclear power get $150 -$300 billion a year in government subsidies worldwide.
- An investment of $660 million would make solar competitive in price – that's 0.5% of the $89,000 million spent by oil companies on exploration in 1998 alone.

What corporations spend

(Based on projections for current and future investments compared with total expenditure for 2001)

Corporation	$ spent on renewables	% of investments
BP Amoco	50 million	3
Shell	100 million	0.1
ExxonMobil	insignificant	insignificant
Chevron Texaco	275 million	2.8

And while oil exploration costs rise, renewable costs are tumbling due to improved technology and economies of scale:

- Wind power costs 10 times less than 20 years ago.
- Solar PV costs 15 times less than 25 years ago.

Worldwatch Institute / NI 357

substitute for 'human judgment' in the interpretation of the scientific evidence.

Debates on the SHEE sciences have educated the public about the limits of objectivity and the roles of values, priorities, choices and exclusions. For a very long time, supporters of 'alternative energy' have pointed to the vast disparity between the meager funds doled out to them for research and development, and the huge sums still lavished on the moribund nuclear power industry.

In medical research, patients' groups have observed how the lion's share of the resources, even those collected and allocated by charities, goes on that 'basic' research which someone claims will solve the fundamental problems of cause and cure of the disease. At the same time, research on the quality of treatments and of care is left on the margins.

In medicine, the reasons are plain: everyone hopes for a 'magic bullet' which will kill the pathogen that makes us sick. Also, that sort of inquiry is useful in building a career in the relevant research science. By contrast, treatment and care are the 'soft', womanly sciences. It doesn't take much imagination to see how particular sets of values and commitments are built into the ruling criteria of quality and the resulting priorities and choices. These determine our knowledge, and our ignorance, in mainstream research science, even when it is nominally dedicated to the objectives of the sciences of safety, health and environment.

Why science is now post-normal

In all these ways, the public are becoming aware that values influence both the shape of what we know, and the selection of what we might know. The old illusion of objectivity, that was fostered by Galileo, is passing into history. We should not reject it completely, for there is a good core of truth there. The task is to distinguish between the routine puzzle-solving science where the old assumptions hold reasonably well, and the areas where a new principle is necessary.

The need for understanding is urgent. In an ever increasing number of policy issues we find science where the uncertainties are gross and the value-commitments are dominant. Looking at issues like global climate change, gender-bending pollutants, the disposal of nuclear wastes, and species extinction, to say nothing of the GRAINN technologies and repro-ductive engineering, we have the shape of the new policy predicaments.

On such issues, no-one can say that science still has the reliability and objectivity that we once took for granted. Now, the facts are uncertain, values in dispute, stakes high and decisions urgent. Indeed, whereas for generations we contrasted hard objective scientific knowledge with soft subjective values, now we have policy decisions that are hard in every way, for which our scientific knowledge is irremediably soft.

Where do we go from here? We can make the transition from the old philosophy of science to the new by showing that it is continuous. There is an intermediate case. Let us look again at the diagram for post-normal science, developed jointly with Silvio O Funtowicz in the course of our long and fruitful collaboration. You see a quadrant, with three circular strips. The ideas behind the names of the axes are 'systems uncertain-ties' and 'decision stakes'.

Down in the lower left hand corner, they are both marked 'low'. That segment is called 'applied science'. This is the 'normal science' where simple puzzle-solving is effective, in the policy-relevant fields of science. All that routine work of monitoring people and the environment, and of building up databases on the effects of environmental factors on the behavior of organisms, comes in this class. It may be unexciting but it is essential.

Indeed, if our safety, health and environment issues

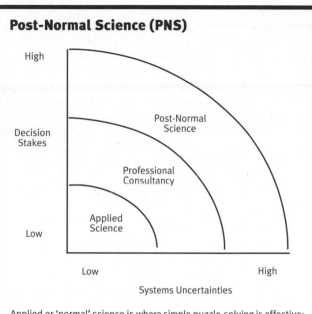

Post-Normal Science (PNS)

Post-Normal Science

Professional Consultancy

Applied Science

Decision Stakes — High / Low

Systems Uncertainties — Low / High

Applied or 'normal' science is where simple puzzle-solving is effective; Professional consultancy is where specialized judgment is needed; Post-normal science is when such professional knowledge and judgment are insufficient. This is when systems uncertainties and decision stakes are high. We cannot assume 'the expert knows best'.

were so much in flux that 'applied science' were ineffective most of the time, then we would be in a very bad way indeed. So far at least, we have not done too bad a job of maintaining the routine work of monitoring, inspecting and regulating the various systems on which our civilization depends. But we know that in many critical cases, especially our major environmental problems, that straightforward 'applied science' is not effective. What next?

What sort of work is involved in that next, intermediate category? We call it 'professional consultancy', in order to suggest the work that is done by a surgeon or a senior engineer. Someone doing those jobs has to be trained in the relevant science, but there is more to the job than just applying the science. Even if most cases are routine and defined by the background science, the professional must always be prepared to use their judgment in coping with the unexpected.

As we all know from the hospital soaps on TV, the surgeon may find something seriously different from expectations when she or he opens up the patient. Similarly, the civil engineer may find difficulties on the site that no survey had revealed. They must not merely be ready to change plans rapidly, perhaps improvising as they go along. More seriously, any mistake can have drastic consequences for their patient or client. The most extreme example of professional consultancy is that of the military commander who knows that there must be a plan before going into battle. But he also knows that on contact with the enemy the plan is the first casualty.

As the problems of technology and environment became more severe through the latter half of the 20th century, the first reaction was to manage them professionally. For example, the design and regulation of the new technology of civil nuclear power called forth a new profession of 'probabilistic risk analysts'.

Its practitioners developed sophisticated mathematical techniques for estimating the probability of various sorts of accidents and the harm that would result from each of them. Their conclusion was that although no-one could guarantee a zero risk, a serious nuclear accident had only a one-in-a-million chance of occurring in any given year. In policy terms this was taken to mean that the technology is safe.

Then came the accident at the Three Mile Island nuclear power station in Pennsylvania, which produced the one-in-a-million type of accident within less than a year after it opened. It was then clear that the risk analysts' profession had not really got the problems under control.

Afterwards, as the more systemic problems of our industrial civilization became acute, we increasingly encountered more such issues, many of them beyond resolution by science alone. We may well ask what sorts of scientific professions are available for managing the emerging problems of the GRAINN technologies that we discussed earlier in this book. That's why post-normal science is now on the agenda.

The insight of post-normal science

Post-normal science reminds us that there are hosts of urgent policy problems involving science, for which routine expertise is totally inadequate, and for which even the best professional knowledge and judgment are insufficient. This happens when, as in the outer strip, either – or both – of systems uncertainties and decision stakes are large. But if all the trained people can't tell us what to do, how are we ever to make good, correct decisions on these difficult and urgent issues?

There is no easy answer. It's most likely that there will be many mistakes, perhaps some of them disastrous. But with the insights of post-normal science we can avoid even worse ones, by refraining from putting our trust in methods that are irrelevant or misleading. We need to recognize and then to move on from the assumption that 'the expert knows best'. She or he might be the researcher or the professional, or even the technician. She or he has the training, and can spout scientific technicalities that leave the lay person totally bemused.

In the conventional model of the process, the expert person starts with the science and then deduces what should be done in practice. This model assumes that the world of practice is sufficiently like the world of science that the expert's deduction is accurate. For 'applied science' it works routinely; for 'professional consultancy', some skill and judgment in interpretation is required. In those traditional cases, people without expert training would seem to have little to contribute to the process of inquiry or decision.

When we come to the situations where post-normal science is appropriate, where uncertainties and value-loadings cannot be denied, we see that the old model of scientific demonstration is inappropriate. Instead we need dialogue. In this situation, everyone has something to learn from everyone else. Of course the experts will have a special command of the more technical issues. But others might know better how well, or how badly, the scientific categories fit in with the reality that they experience. Many policy debates hinge on 'safe limits'. It doesn't require a PhD to be able to ask intelligent questions about safety tests and to inquire whether they are truly realistic in relation to practice.

We might query whether lab tests or surveys, even if performed quite properly by 'applied science' criteria, might turn out to be irrelevant or misleading if applied uncritically in a post-normal situation. Thus, we might need to know whether the sample populations in a study included, for example, children and pregnant women, or animals that breathe air close to the soil. Or we might need to know whether the specifications for safe use of equipment are likely to be respected in real industrial or agricultural situations – in some locations it may be prudent to assume that they will not. Up to about a decade ago, government regulators did not recognize that children have significantly greater susceptibility than adults to environmental hazards.[2] Before then, how 'objective' was the official SHEE science in this regard?

All such observations and criticisms can be put by people who have independence and common sense. People with local or practical knowledge can spot these sorts of flaws more effectively than experts who are trained up in a standard doctrine, applicable to a textbook world.

The 'objectivity' that relies on an absence of people, judgments and values is no longer appropriate as an ideal. Instead, we should cultivate 'integrity' in science. For our dialogue on policy issues, we need participants to engage in a 'negotiation in good faith'. Each advances their case on the basis of their own clear and open perspectives and commitments. All participants recognize their uncertainties and areas of ignorance, and respect the integrity of those with whom they disagree. The questions, 'What if ...' and 'What about ...?' can be appreciated as offering an essential stimulus to the dialogue. It all amounts to reminding us all of Murphy's Law – the notion that anything that can go wrong will go wrong. This is something that is, as yet, totally absent from scientific

The rabbi's dilemma

An inspiring example of the creative power of an Extended Peer Community is provided by the campaign of an Orthodox Rabbi in New York to eradicate Tay-Sachs disease from his community. This devastating disease of childhood is caused by a recessive gene found among Ashkenazi (European) Jews. It is more prevalent among the small, inbred Orthodox communities; and Rabbi Josef Eckstein had repeated experience of the tragedy. Attempts to screen and counsel couples are blocked by the stigma of being a carrier. Rabbi Eckstein used an insight that comes more naturally to a rabbi than to a geneticist. The key to his system of recording and checking is the right of a possible carrier of the gene to be ignorant of their condition. Couples are told of the risk only if both are carriers of the gene; otherwise the genetic information is kept in a confidential database.

After the usual uphill struggle, he won acceptance from the medical establishment. Now Tay-Sachs (along with other genetic diseases) is effectively eliminated in that population and in Israel as well. And with his, rigorously anonymized, database he has been able to contribute, indirectly and directly, to research on various genetic diseases. ∎

Alison George, 'The rabbi's dilemma', *New Scientist*, 14 February 2004 training, but which is totally necessary for good policy decisions in the real world.

Appreciating the vital role of those others in the dialogue, we call them the 'extended peer community'. For they are full participants in the process, learning and also teaching in the dialogue. And they bring with them what we might call 'extended facts'. Scientists will necessarily and justifiably focus on the information that is produced under the standardized, idealized conditions that are necessary for successful research. But the extended peer community has other sources.

In policy issues, investigative journalism is a key resource, along with documents that were not originally intended for public view. In addition, there is local knowledge, including the place, its inhabitants of all sorts and species, and its history, traditions and special or sacred values. All these 'extended facts' are vital to the policy processes. They are excluded from the perspective of the 'normal' experts and professionals; it is the post-normal extended peer community that introduces them as valid contributions to the debate (see box, 'The rabbi's dilemma'). The two peer communities, expert and extended, are fully complementary in post-normal science.

What's in a name?

As we consider the essential role of the extended peer community, our vision of post-normal science reminds us of a great variety of endeavors to adapt science to the needs of a modern democratic society. People have spoken of 'critical science', 'citizens' science', 'civic science', 'community research', 'action research', 'open science' and 'see-through science', as well as 'environmental', 'ecological' and 'sustainability' science.

Each title has its own flavor, and its own authentic perspective on the whole problem. We offer 'post-normal' as a member of that family. For us, it expresses two key insights. First, that these times are far from 'normal'. Second, that 'normal, puzzle-solving science is now totally inadequate as a method and a perspective, for the great policy issues of our time. Uncertainty now rules political as well as environmental affairs. And the value-commitments of people, reflected in their lifestyle choices, will make a big difference to the chances that the human race eventually makes it through to sustainability. In very many fields of public policy formation and decision making in advanced societies, 'civil society' is included in the processes. Post-normal science can be understood as a recommendation that this be done even in cases where the core of the issue is scientific.

This new perspective on science has benefits for the scientists themselves. They can be liberated from the confusion and self-doubt resulting from their discovery that some scientific problems cannot be solved by 'normal' methods. The failure to produce conclusive information about pollution or climate change is not the fault of the science or the scientists themselves. It is because we live in a new age of policy where science is necessary but not sufficient for solutions.

For their part, the extended peer community are no longer relegated to second-class status, and their special knowledge is no longer dismissed as inferior or bogus. They are full partners in the dialogue, who have much to teach as well as to learn. Both sides benefit from the dispelling of the illusion of scientific objectivity. That is the way forward, as expressed in the title 'post-normal'.

1 Jerome R Ravetz, *Scientific Knowledge and its Social Problems* (Oxford University Press 1971; Transaction Books, New Brunswick, New Jersey, 1996). 2 Sandra Hoffman and Alan J Krupnick, 'Valuing risk to health', in Resources (Resources for the Future) 154, Summer 2004.

6 Uncertainty

The great achievements of science of past centuries appeared to conquer a capricious and uncontrollable natural world. But all that has changed and our uncertainties about the human-made world are now severe. To be aware of our ignorance may provide a key to creating an appropriate science for the future.

ALONG WITH 'OBJECTIVITY', the other great falling idol of contemporary science is 'certainty'. This is an even steeper descent, for certainty has been the hallmark of genuine science, for teachers and for propagandists, for a very long time indeed. And of course there is a large core of science whose certainty is not in doubt, at least not in the short run.

But when we leave the realm of that 'normal' science and go out to where the real challenges are, then we find that certainty has been left far behind. The traditional teaching and propaganda of science have given us very little preparation for this new state of affairs; and so it is vitally important for us to get clarity about it.

Uncertainty is not just a black hole of disbelief. It is a special state of knowing, whereby imperfections in our knowledge can be recognized, understood and used creatively. Without making a fuss about it, working scientists have always been managing uncertainty in order to get through their work. It is only now that the philosophers are beginning to catch up with them and are starting to analyze their successful past practice.

We know that many people are still shocked by the sight of scientists disagreeing in public. Perhaps the very idea of there being uncertainty in science is too much for them to take. After all, if the scientists are disagreeing then one side must be wrong! And if scientists can be wrong, then what security is left? Some scientists imagine that the public would panic at this discovery and so they believe that such disagreements are like arguments between parents, not to be done in front of the children.

But perhaps the public reaction is more sophisticated. It's not the disagreement that is unsettling, but rather the sight of scientists who themselves don't know how to handle the uncertainty. The scientists in the debate must have gone through this sort of thing before and yet they seem unable to imagine that this time they themselves might be the ones who are wrong. That doesn't say much for their understanding of themselves, or indeed of science.

Such dogmatic scientists are the victims of their indoctrination. The science of the textbooks and exams offers perfect certainty. For years, perhaps for many years, science students never see a problem that lacks a single neat correct solution. Those students who feel uncomfortable or alienated by this artificial certainty tend to just get out of science. When the real world suddenly throws up problems that have many possible solutions or none, and the simple certainties of science are lost, it's no wonder that the scientists might feel confused and betrayed. They might well retreat into an extreme dogmatism about scientific knowledge in general and about their own research in particular. The public occasionally gets a glimpse of quite breathtaking arrogance, as when some cosmologists claimed to know for certain what had happened in the first 10^{-35} of a second of the existence of the universe.

Some fortunate scientists get personal guidance when they start research. In a largely informal way they are taught the craft skills of managing the various sorts of uncertainty in their work. Then if they have occasion to reflect on the limitations of their special knowledge, they have some sort of personal security based on their successful practice.

In this chapter I will provide an analysis of uncertainty that expresses the practical understanding of those reflective scientists. It will enable us to go beyond disillusion over the lost ideal of perfect certainty in science.

The sources of uncertainty

I've already mentioned that when two groups of scientists disagree, certainty is lost. They can't both be right, but someone looking at the debate from the outside can't easily determine which side is wrong. An even more serious case of loss of certainty is when the scientists don't disagree, and then things turn out to be different from what they had all announced. The public sees this most dramatically when there is an official consensus that something is safe and then it turns out to be dangerous.

Of course, there are always dissident voices on any issue and most of the time they are wrong, at least in the short run. The dissidents comfort themselves with the reminder that Galileo was in the critical minority in his time and he turned out to be right. And every now and then the isolated critics are indeed proven right. This is more serious, for then the public senses that something is seriously wrong about the way that the mainstream scientists engage with opposing viewpoints. Also, we are left wondering how many other universally accepted scientific assurances of safety are wrong – and when we will find out about them too.

Well, how can things go wrong in science? Of course, there will always be new discoveries, that could not have been anticipated in advance. These make everyone think again about their theories. That's the nice way for things to change. In the policy domain, it's more common for there to be a mismatch between the elegant simplicities of a theory cast in scientific terms and the messy complications and complexities of a real situation that needs to be managed.

There's a history here that we need to appreciate if we are to comprehend and resolve our present difficulties.

The ideas of progress and certainty in science are deeply embedded in the way we see the world, and justifiably so. For a very long time indeed, our uncertainty and indeed our ignorance about ourselves and our environment was very great. We were at the mercy of what seemed to be a capricious natural world.

It was understandable that people should believe that some super-intelligences were running it all, and so they would try to placate or please those invisible beings with rituals and magic. Of course, through all those ages of confusion, humanity's control over the natural world evolved and improved. This could not have happened without increased understanding as well. Looking at the accomplishments of earlier civilizations, we need to speak not only of engineering, but also of science.

Over the centuries the areas of scientific certainty and technical power grew, gradually squeezing out those controlled by magic and religion. In modern Europe, the process accelerated. At the beginning of the 17th century, astrology and the stationary earth were both common sense. By the end of the century, astrology was discredited and the earth floated through the heavens.

During this era of progress, miracles and magic steadily dwindled in importance. Some of the most important prodigious phenomena were explained away. The rainbow, the Biblical symbol of hope, was shown by Descartes to result from the refraction of sunlight in raindrops. Lightning – or 'thunderbolts' – was revealed by Benjamin Franklin to be just a huge electric spark. Lightning-rods protected churches better than prayers ever did. It really became plausible to believe that with the aid of science we would soon conquer disease, poverty, ignorance and war. Visionaries proclaimed that humanity could make its own paradise on earth. Science, the vehicle of progress, was absolutely certain, and perfectly beneficial. Only with World War One and the use of science for the first 'inhumane' weapon, poison gas, did serious doubts emerge.

When things went wrong in the later 20th century, it was not merely the invention of supremely dangerous weapons like the atomic bomb. Worse, there were the unanticipated consequences of science-based industrial processes, first made notorious in Rachel Carson's *Silent Spring*. Sometimes these could be easily identified and remedied, as they were in that particular case. But more frequently, the pathways to harm have become so complicated, and the necessary societal responses have become so complex, that traditional science alone simply isn't up to the job.

The problems of waste, a subject that could seem too boring and disgusting to merit scientific study, can become crucial in determining our survival. Our uncertainties have become severe and our ignorance may turn out to be lethal. The simple optimism of earlier ages has vanished.

This is not to say that we should abandon science; not at all. But there are two important lessons to learn from the failure to conquer uncertainty. The obvious one is that we need a new conception of science, one based on coping with uncertainty rather than pretending to be achieving perfect certainty.

The deeper lesson is that when we engage with any new development in science or technology, we should assess what sort of knowledge we will need, in order to keep it safe. And then we should get some idea of where the critical uncertainties lie and what decisions we should make in the light of our knowledge of them. A start on this has been made by the Netherlands environment agency RIVM, with a 'Guidance' on managing uncertainty for scientists.[1]

When we look back at the magic and superstition of the earlier ages, we see that there might have been an important truth there. This was not necessarily about the enchanted world that people believed in. Rather, it

Long-lived radioactive wastes

A potentially disastrous failure of scientific certainty has occurred with long-lived radioactive wastes. These are an inevitable consequence of nuclear weaponry and civil nuclear power. They are poisonous for thousands, in some cases many thousands, of years. At the start of military and civil nuclear programs, problems of waste disposal were of no great concern. Since no scientist had ever gained a reputation with theories about garbage, radioactive waste management was a low-prestige, low-priority affair. It was just assumed the most dangerous of them – not very great in bulk – could be embedded in some sort of solid medium and stored out of harm's way.

But in the prophetic words of environmental biologist Barry Commoner, 'everything has to go somewhere'. And the 'somewhere' turned out to be very problematic. Any encasing medium turned out to be vulnerable to damage caused by the radiation being emitted by the waste. Also, any subterranean location turned out to be vulnerable to leaks.

Eventually the US Federal Government ruled that the dangerous wastes would be stored under Yucca Mountain in Nevada. They have expended enormous sums on computer simulations to establish that this site would indeed be safe. But these simulations are only as good as the data that goes into them and it isn't easy to get good data about possible cracks in rocks deep underground. Nor is it easy to make accurate models of the behavior of inaccessible complex rock strata.

It was easy for the opponents of that storage site to expose the gross uncertainties in the government's computer models. And in the meantime the waste from civil nuclear reactors accumulates ever more dangerously, packed ever more tightly in the cooling ponds of power stations all over the US. The military waste – administratively, a completely separate problem – becomes a radioactive hell on earth, ever more menacing as its containers become more aged and corroded. In the face of these ineradicable horrors, science is – as yet – ignorant and impotent. ■

Kristin Shrader-Frechette, *Risk Estimation and Expert Judgment: The Case of Yucca Mountain* hhtp:www.fplc.edu/RISK/vol3/fall/Shraderf.htm

was about ourselves and our place in the world. When science became all-conquering, some thought that men would become like gods. That was a serious mistake; the world is far bigger than ourselves. As the old saying goes, our pride has been leading us to a fall.

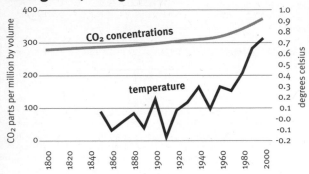

The heat is on

To stabilize the global climate will require huge cuts in greenhouse-gas emissions – and a big increase in justice. The UN-endorsed Intergovernmental Panel on Climate Change (IPCC) has stated that emissions reductions of 60% by 2050 are essential. But many experts now think even this is not enough – 70 or 80% is needed.

Rising heat, rising carbon

In 2002 the British Meteorological Office's Hadley Centre reported that climate change was happening 50% faster than originally believed. By 2040 they predict most of the world's forests will begin to die. ■

The previous centuries of steady scientific progress had made a safe and secure life 'normal' for at least the fortunate minority of humanity. Our discovery of crucial and troubling scientific uncertainties has gone in parallel with the end of 'normality' in our ordinary lives. Global climate change and global terrorism are two facets of a new situation of deep uncertainty and danger.

Managing uncertainty

In their ordinary work, most scientists, either students, teachers or researchers, do not need to cope with this big picture of the loss of certainty. There is a 'normal'

Current and historical sea level rise in selected island countries

Country	Average sea level rise, 2002 (millimeters)	Long-term sea level rise (millimeters per year)
Cook Islands	12	2.3
Fiji	2	4.0
French Polynesia	24	2.5
Galapagos	52	1.5
Japan	6	3.2
Kiribati	35	−0.2
Maldives	8	–
Saipan	6	–
Seychelles	6	–
Tonga	40	4.9
Tuvalu	38	0.9

University of Hawaii, Permanent Service for Mean Sea Level, and the South Pacific Sea Level and Climate Monitoring Project

uncertainty that is managed by the 'error bars' that every science student learns about. These are the thick vertical lines associated with data points on graphs that display the 'spread' of the numbers reported there.

The numbers derived from successive tries of the same experiment always wander a bit – sometimes a lot. There is no way to know which is the 'best' number. So scientists have mathematical methods for assigning the best 'representative' number, and also for indicating the sort of uncertainty that lies around it. Out of these calculations come the 'best' number and also the uncertainty interval around it, indicated by the 'error bar'.

For most of 'normal' science, that's quite enough. But even in research science, there is more to the story of uncertainty than simple error bars. Few measurements – outside of the elementary science lab – relate simply to 'the thing itself', like length. Most are highly indirect. We have already mentioned the need for judgment in the management of 'outlier' data. Also, in any real experiment, the equipment needs first to be calibrated and then needs to be checked regularly to confirm that the calibration has not drifted.

There are all sorts of sources of error in the handling of experimental samples. It needs to be confirmed that they are as pure as they are claimed to be. The instrument needs checking to ensure that it is measuring what it is believed to be measuring and not recording some 'artifact' instead. In statistical surveys, there are pitfalls everywhere, as in the possibility of incorrect sampling, incompetent samplers, faulty data-processing, data sets that do not fit the assumptions of the significance test and so on.

All this uncertainty might seem strange to someone raised on the apparent certainties of the textbooks. The student might ask, scientists discover facts, don't they? How could mistakes occur? Well, we have just seen that measurements are uncertain, and it's easy to imagine that theories are also incomplete and uncertain. In addition, scientific arguments are not logical deductions like those of mathematics. For many years philosophers tried to articulate 'The Scientific Method' by which scientists always get it right. They failed.

Arguments in science are compounded of inferences of different sorts, none of them valid in itself. When we 'confirm' a theory by a successful experiment, that is a standard fallacy, called 'affirming the consequent'. It has been known to logicians for many centuries. In a caricatured form, we can argue: 'If the moon is made of mouldy cheese, it will have spots'; 'The moon has spots'; 'Therefore the moon is made of mouldy cheese'. Wrong and ridiculous; but that's the logic of deduction and confirmation.

The other main stream of scientific argument is 'induction': if event B happens every time we produce A, then A must be the cause of B. If life were so simple, then we wouldn't need all those sophisticated statistical tests. There are many pitfalls in that inference: the correlations may themselves be temporary or spurious, or A and B may both be caused by another factor C.

Even in mathematics, which is genuinely logical, constructing a proof is rather like writing software: trying an argument, and hoping that you have found all the gaps. What happens in scientific argument is that a mixture of patterns of inference, including deduction, induction and analogy, are employed in a largely informal fashion. All the known pitfalls in inferences are bridged over by precautions and corroborations. The arguments are never absolutely certain, but good craftsmanship can make them robust. (See my earlier book *Scientific Knowledge and its Social Problems* for a discussion of these methodological issues involving pitfalls).

Even when all goes well, in the best run laboratories, improvements in instrumentation and technique can force corrections and revisions in previously accepted results. The most basic of physical constants can experience a 'bouncing' effect over time, as the graph shows. (See Fine-Structure Constant box, next page).

It is most remarkable that this 'bouncing' might introduce more uncertainty than the error bar itself. A look at the graph confirms this. At least half the changes from one 'accepted' value to the next were greater than the associated error bar. Of course, the constant finally settled down and became nearly constant; perhaps by then it was not so interesting. So in addition to the 'inaccuracy' of a result we need to describe another sort of uncertainty; let us call it 'unreliability'.

To express this, scientists sometimes cite an item of data with two ± terms. We might see $5.3 \pm 2.1 \pm 3$. The first comes from the error-bar calculation, and the second from their assessment of the history (and future) of the number. We see again how the supposedly objective quantitative information needs to be qualified by skilled personal judgments, if its information is to have genuine content.

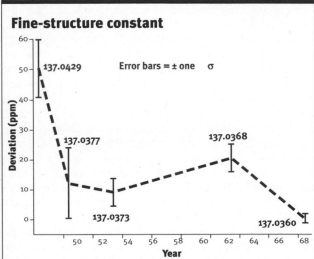

Fine-structure constant

Error bars = ± one σ

137.0429

137.0377

137.0373

137.0368

137.0360

Deviation (ppm)

Year

The 'fine-structure constant' is a fundamental building-block
of physical theories of atomic structure. This graph shows
the successive 'recommended values' of the constant, during
the years from 1950 to 1968. These values are published by a
committee of experts, who assess all the experimental reports
that come in during the year and decide which are the most
reliable.

For each recommended value they add an 'error bar', denoted
by a vertical line. This indicates the range where the true value
of the constant is expected to lie. As experimental techniques
improve, these 'recommended values' change, and so the constant
will inevitably 'bounce' until it settles down. But sometimes the
'bounce' is greater than the error bar.

That is, the scientists are frequently surprised when they come
back to a renewed assessment of the 'recommended value'.

This sequence of uncertainties may begin to remind
us of the doctrine of post-normal science. First we had
the routine puzzle-solving, where all the uncertainty
can be tamed by calculations, expressed by its error-
bars. That related to the unavoidable inaccuracies in
our basic data.

Next in the post-normal scheme came the greater uncertainties of 'professional consultancy', combined with the greater decision stakes. Corresponding to that situation we have unreliability of the theories and the assumptions that provide the framework for our observations and measurements. To manage these, we make judgments, where the mathematical calculations are supplemented by experience.

Is there another sort of uncertainty, that becomes crucial when we are in a post-normal situation? Yes, and its name is ignorance.

Ignorance and science

How could ignorance have anything to do with science? After all, the great goal of science is the conquest of ignorance. Once upon a time people believed that that final conquest was imminent; but now we know better.

As we saw when discussing the consequences of the GRAINN technologies, any decision on policy for science and technology must cope with our ignorance about the future. We cannot predict, and so we must – if we are to be sane – adopt precaution.

To manage ignorance properly will require quite a different mindset from those which are successful for textbook learning or puzzle-solving research. We can see the difference in terms of the sorts of characteristic questions that are asked. In ordinary research or development, the inquiry proceeds by converging in on quite specific questions. We might want to know the composition or structure of some substance, or to understand the causes of some reaction. In development, we have a device in mind, that we want to design and create, that should work under specified conditions. In very abbreviated form, we can describe such questions as 'what/how?' and 'how/why?'.

But when we are dealing with issues in the SHEE – safety, health, environment – sciences, and are aware of M&M – malevolence and muddle – and of our ignorance, we have to ask some quite new questions. Such questions may take the form of 'what-about?' and 'what-if?'. These are the sorts of questions that typically come from people who have not been trained – and constrained – in the standard doctrine of the relevant expertise. They are 'divergent' questions, inviting exploration into the unknown. They are also precautionary questions, serving as reminders that we do not know everything about the situation – and might not know as much as we think we do.

If we reflect for a moment on the possibility that our knowledge is not so extensive or so secure as we believe it is, we can be led into strange domains of thought. For this involves another sort of ignorance. This new sort does not concern our knowledge, or lack of it, of what is out there in the real world. Rather, it concerns our knowledge – or our awareness or lack of it – about our own knowledge and ignorance.

To give a simple example: how many times have we read or seen something about some unfamiliar scene or activity and exclaimed: 'I never imagined that such things could be'. Those things might be some strange objects or processes in the natural world, or some strange variety of human behavior or belief. That sort of deep discovery is fairly common among students, when they leave the sheltered, restricted environment of home and school, and encounter all sorts of new experiences at university or college. Needless to say, such experiences can be very maturing, but also very unsettling in the short term.

One way of describing such experiences is to say that our imagination has been expanded, not merely in quantity but in quality. We have learned many more things, but we have also learned that there are things out there that we had never imagined. Another way of putting it is that we have discovered our previous ignorance. We had not been aware of our ignorance of all those strange and new things; speaking paradoxically, we had been ignorant of our ignorance.

It is well worth while the reader's time and effort to go over those two last paragraphs, for they convey a most important message. This is, that one of the main functions of education is to make us aware of our ignorance. This was what the great philosopher Socrates was all about; he would play verbal games with his friends, to shake them out of the cage of their complacency and bring them to a realization of their ignorance.

Curing ignorance of ignorance is not merely a concern of enlightened educators. It is vitally important for science and survival. For some centuries, philosophers and science teachers have together fostered an ignorance of our ignorance.

Who has ever seen a discussion of ignorance in a science course, or seen a question about ignorance on a science exam? By putting all the emphasis on what we know and suppressing our awareness of ignorance, that tradition is hampering our efforts to guide science to a more prudent and sustainable future.

But in this area of the understanding of science, as in others, things are changing rapidly for the better.

In big policy issues like climate change, ignorance is invoked by both sides, each advocating the sort of 'precautionary principle' that suits it best. One side favors the ecosystem and the other favors economic growth. In this way, ignorance has been made natural and comprehensible. In these debates, a new principle has emerged: 'It's what you don't know that you don't know, that you need to worry about'.

For some, this is obscure nonsense. For others, it expresses the core of the problem of sustainability and survival of our modern science-based civilization. When that paradoxical statement becomes common-sense, then we will be able to say that our culture has fully rediscovered the awareness of ignorance, and then is ready to move forward with an appropriate sort of science.

Ignorance, precaution and post-normal science

We are now in a position to see what goes wrong in modern technology, and how it might be put right. First, from Murphy's Law we know that every innovation will eventually have some adverse consequences. Predictions about these will necessarily be speculative – and mostly wrong – while the technology is still embryonic. At that stage, ignorance still dominates. However, as the technology matures, predictions can become more precise, although still uncertain in varying degrees.

In some areas of obvious risk, such as medicines, 'early warning' systems are in place; but elsewhere they are patchy or nonexistent. Hence when harmful effects first turn up, they can easily be missed. The earliest reports will be uncertain or ambiguous. They can be dismissed as 'merely anecdotal evidence'. In terms of the Post-normal science diagram (see chapter 5), they are far out on the horizontal axis, with high uncertainties and relatively small decision stakes.

Those responsible for safety research must then decide on how strongly to investigate the warnings. Science alone does not have the answer; it depends on the assessment of the relative risks, just as in the management of 'outlier' data. If your interest is likely to suffer from the harmful effects, you will be strong on 'precaution'; but if your interest is likely to suffer from their confirmation – resulting in loss of sales or reputation – you will be less strong.

Up to very recently, the balance of risks has favored the vested interests producing or promoting innovation, that is industry and government. Government regulatory agencies do not function in a vacuum and so they can be pressured, even to the point of corruption, in favoring innovation over safety. In post-normal science terms, even as the systems uncertainties decrease, the decision stakes increase – and the issue moves towards a position far out on the vertical axis.

Anecdotal evidence can be the victim of a 'Catch-22' effect. The authorities can refuse to consider it because it is 'merely anecdotal', and then use the same excuse to refuse to fund the research that could make it stronger. For a doleful tale of real warnings that were systematically ignored for decades, see the report *Late lessons from early warnings: the precautionary principle 1896-2000* European Environment Agency.[2] Here we find all the classic stories – of asbestos, PCBs, halocarbons, DES and BSE, together with others equally instructive, and a set of 12 recommendations for practice.

Now, however, a more sophisticated and alert public has shifted the balance of power and risk. The exposure of a cover-up – usually involving an 'extended peer community' of whistle-blowers and investigative journalists – can impose severe costs for the perpetrators, both in loss of reputation and sales and also in vulnerability to direct-action campaigns and legal proceedings. The more enlightened voices within industry and government now have practical arguments to support their commitment.

A 'precautionary principle' favoring safety is becoming good business sense as well as serving the cause of a just and sustainable society. Ignorance and uncertainty can be managed as part of a dialogue on appropriate balances among the various perspectives and interests in a post-normal science issue. They can no longer be simply manipulated for the furtherance of profit and power.

1 See www.nusap.net 2 European Environment Agency, 'Late lessons from early warnings: the precautionary principle 1896-2000', *Environmental Issue Report No 22*, Copenhagen, 2001.

7 Science and democracy

Historically science has been associated with the advance of democracy. The correlation was not always perfect. Nor is it today. And in poorer countries science can easily become part of the democratic deficit rather than a benefit.

TWO OF THE greatest positive themes of our age are science and democracy. We now tend to judge institutions and societies by the criterion of democracy; and it is universally accepted that science is the foundation of the economic growth that enables modern democracy to flourish. Because the relationship is so important, we need to scrutinize it closely.

But first we must resolve a paradox. Within the discourse of science as presented in textbooks, there is no mention whatever of democracy. From one point of view, any mention of 'democracy' in relation to science might seem to be mixing totally different things. Can a scientific fact be more or less democratic? It might just as well be more or less brown!

I have already argued that the science syllabus should include more material about uncertainty and values, but politics is surely in a totally different realm. People can study science, and indeed pursue a career in science, with great success while knowing or caring little about the state of the society around them.

Churches and dictators

When we take a broader perspective on science, it turns out that democracy is relevant, and in a variety of ways. First, there is a long history, extending over some centuries nearly to the present, when 'science' was enlisted in the struggle against the domination of peoples' thinking by theology and the domination of social life by religious institutions.

The issues were never simple. But the broad lines of conflict had scientific progress, with liberty and eventually democracy, on one side, and institutional religion, based on obedience and ignorance, on the other. The last big battle in this long war took place in Central Europe in the 1920s and 1930s. The 'logical positivist' philosophy of science, which held sway in the US until the 1960s and beyond, was one of its products.

There is another, more recent connection between science and democracy. This grew out of the experience of the dictatorships of the 20th century. For those on the side of democracy, it seemed obvious that dictatorships would be ultimately self-defeating, as they would necessarily destroy science and thereby weaken themselves fatally. Science depends on freedom of inquiry and fairness of procedure, both of which are impossible under dictatorships.

This view was confirmed by the experience of Nazi Germany, where Jewish and anti-Nazi scientists were expelled or, later, murdered, and of Soviet Russia, where nonconforming scientists were exiled, imprisoned or 'liquidated'. The failure of the Germans to create an atomic bomb was further evidence for this theory. When the Soviets later did just that, their success was – in part justifiably– explained away as the result of their being given 'the secret' by spies.

However, it was eventually realized that while some parts of science did indeed suffer badly under those dictatorships, others could do surprisingly well. Social medicine and public health did so, even under the Nazis before the war. Later, space technology flourished under the Soviets. So while the connection is real, it is far from absolute.

Democracy and the world of research
There is yet another aspect of science and democracy that is worth mentioning. In some respects the world

of research is the most egalitarian possible. Papers for publication are accepted on merit, and scrutinized for their scientific quality. The social status, race, creed, gender or sexual preference of their authors are totally irrelevant to the reception of their publications. If the results are important, they are taken seriously, and if not, not.

Of course, this picture is somewhat idealized. Belonging to a prestigious institution or having a prestigious co-author can help in getting publication in a prestigious journal. But it is true that in science the outsider stands a better chance of being heard than in most other areas of social activity. In that respect at least, science does embody democracy.

However, when we look at the working life of scientists, the picture of democracy becomes somewhat clouded. In the first place, to become a scientist you need to get on the ladder at an early age. Up to about a century ago, it was possible for someone to teach themselves enough to make a successful career in science or invention. In the US, people like Thomas A Edison, George Washington Carver and Luther Burbank became real folk heroes for the way they started from nowhere and became great scientists or inventors. But by now that path has become just about impossible to follow.

We might say, paradoxically, that a minority person from a ghetto has a better chance of becoming President of the United States than of becoming a leading scientist. If you're not on the right rung of the ladder at the right age, you just won't get there at all. We have also discovered that women are at a systematic disadvantage in science. This is partly due to traditional gender prejudice, which until quite recently was just as vicious in science as anywhere else; but it is also a result of the current nature of scientific careers which can be fatally interrupted by taking time out to have children.

The conclusion of this analysis shouldn't be too depressing. Yes, science is broadly democratic, certainly more democratic in its daily operations than, say, business. But it is still not too far ahead of the rest of society and it has its own sorts of imperfections.

In this discussion we are considering 'science' as research or development, conducted in its special institutions. When we look at science out in the world of policy, and as experienced by people, the whole question of democracy takes on a different shape and also greater practical importance.

Science to better the world

For a long time people who wanted to improve the world through science had an idealistic, perhaps even simplistic, picture of the process. First, some enlightened people would realize that there is a serious social problem out there. It might be poverty, disease, or some special social pathology. But the established authorities generally didn't want to know.

Doing something about such problems requires resources, and might also offend some special interest. Science is then brought in to convince the public and rally its support. The facts are gathered and marshaled in an argument that shows first that the problem is real and second that it can be solved.

There were some great studies, mainly of poverty and its effects, through the Victorian period, which established this tradition. The scientists who conducted this work saw themselves as standing outside the ordinary run of politics. They could not be corrupted by the rewards of power. Their only allegiance was to the truth.

Not everyone shared their belief in their purity, for they frequently had their own special agendas for the solution of difficult societal problems. Right up to World War Two, some leading biological scientists believed that 'inferior' sorts of humans could be identified by 'eugenic' science. They could then be prevented from adulterating the stock of the human race by being discouraged from breeding, for example.

Such science was totally discredited by the Nazis' racial theories that culminated in the Holocaust. Even medical science was infected by racism, as when black patients in the Southern states of America were allowed to sicken and die of syphilis, for the sake of a complete medical record. The sad story of science put to the service of racism has been told by Steven J Gould in *The Mismeasure of Man*.[1]

Nowadays we are all more sophisticated, even a bit cynical sometimes. When we read about some policy-relevant research, we immediately ask: who is sponsoring it? And then all too often there is the dismissive judgment: 'They would say that, wouldn't they?' This can be quite unfair to the scientists performing the work. Someone might quite honestly and legitimately arrive at a certain perspective on the problem, and

---

Not on the packet

Tobacco firms have a reputation for manipulating (and hiding) unfavorable scientific research. This might include the fact that tobacco smoke is a potent mix of over 4,000 chemicals.

It includes:	as found in:
Acetone	paint stripper
Ammonia	floor cleaner
Arsenic	insect poison
Butane	lighter fuel
Cadmium	car batteries
Carbon Monoxide	car exhaust fumes
DDT	insecticide
Hydrogen cyanide	gas chambers
Methanol	rocket fuel
Naphthalene	moth balls
Toluene	industrial solvent
Vinyl chloride	plastics

The Tobacco Atlas by Judith Mackay and Michael Eriksen, 2002, WHO, Geneva / NI 369

then find that her or his only source of funding is an institution for whom that perspective is useful. So the scientist takes the funding, but keeps the commitment to publish what she or he discovers, regardless of the outcome. Unfortunately, some institutions, notably the tobacco firms, have such a bad reputation of manipulating research to their own ends, that anyone who takes their money is branded as either hopelessly naive or dangerously opportunist.

How do such issues relate to science and democracy? After all, in a free society anyone can sponsor research and take their chances in the public debate.

Unfortunately, it's not so simple. It's now hard to deny that even when the structures of democracy are in place, undue control of the information reaching the public can hamper and deform the democratic processes. In the United States, some say that 'big business' controls the media, while others say the same about 'the liberals'; but they both implicitly agree that monopoly control of the media is bad for democracy.

The same concerns apply to science that is related to policy issues. These issues are frequently complex, value-laden and uncertain. If only one side of the scientific debate reaches the public, then public opinion will be misled and liable to make the wrong decisions.

Science in policy issues

We can get a better understanding of policy-related science and the public in a democracy by analyzing the sequence of the processes whereby the science is produced. Although the core of the work consists of acts of discovery, which cannot be predicted, that core is defined by many choices. These reflect prior commitments. Moreover, they can strongly influence what is eventually discovered and what is not. In this way, our knowledge and our ignorance are influenced by the values and commitments that underlie the choices that select and shape the research that is eventually done.

First in the sequence, there is the definition of the policy issue itself. Although the headlines might make them seem simple, issues like 'reducing accidents' or 'stopping pollution' or 'fighting disease' are far from simple in practice. There will always be contrasting, frequently conflicting, perspectives on what the issue actually is. For example, if we are worried about road accidents in a particular place, then we have to consider the drivers, the passengers and the other road users – cyclists and pedestrians of different sorts – all as people 'at risk' in different ways.

For managing the risk we have the designers, engineers, vehicle manufacturers, regulation and administration at various levels, plus those who make and enforce the relevant laws, as well as the behavior of all the different users. Beyond them we will have the general cultural atmosphere, the attitudes to the proper uses of roads, to legislation that is restrictive or intrusive and finally to the levels of risk to various affected people that are 'acceptable'.

For each such 'stakeholder', however defined, this particular problem of risks will be conceived and evaluated through their broader understanding of themselves and their concerns. As a very simple example, one can reduce injuries to persons inside a car once an accident has happened, by designing the car to survive impacts better. But that policy is liable to increase injuries to those outside who are in a smaller vehicle or in none at all.

Alternatively, the authorities can reduce pedestrian accidents at a 'black spot' by preventing them crossing there by means of physical barriers. But they are liable to find that accidents increase at the next crossing up the road, and also that some impatient people will jump over the barriers to cross at their favorite spot, breaking the law and causing even worse accidents.

We are far removed from having a simple policy problem whose solution is determined by the simple facts of science.

If we set a scientific problem of 'reducing risk' or even of measuring it, we must first define the policy issue in which it has meaning. Designing a 'safer' car will not necessarily help those who are liable to be hit by it. In this way, the science is subsidiary to the policy. Scientific research that does not have a high priority, in policy terms, will have less of a chance of being done. As a result, that bit of knowledge, which might be of importance to some stakeholders, will not exist. They will remain in a state of ignorance and to some extent that ignorance has been created by policy. The debate will then be skewed in favor of those who dominate the prior definition of the policy problem.

The science that is done may well be objective and accurate. But when we consider the science that is *not* done, we get a better idea of how policy, and hence politics, influences the knowledge that we have and hence the decisions that we later take. It does not take much political wisdom to realize that where a society is less democratic, the issues, the science, the knowledge and the decisions will all favor the vested interests at the expense of those without the power. Thus the awareness of our ignorance in policy-related science is not merely a philosophical issue, but is one that has direct consequences for our understanding of how science works in a democratic society.

A similar influence of society on science occurs in the methods adopted in a scientific inquiry. In our discussion of uncertainty, we showed that scientific arguments cannot be conclusive. Whatever might be the logic of science, it cannot produce indubitable truth. In practice, arguments in science become more like arguments in the law than we had hitherto realized. In particular, all science depends on an implicit assignment of burden of proof in investigations (see box). In 'normal' science this is usually accomplished by an informal consensus of the peer community, and most researchers never even realize that this is a policy decision that might have been otherwise.

When we come to policy-related research, the choice of burden of proof can be crucial; and this is a prior policy, or political, decision. If there is evidence of harm, but not quite of sufficient strength to pass the stringent tests of significance that are appropriate for lab science, do we count it as a 'fact' worth publishing? If we reject it, it is forever buried in someone's lab notes. In that case, the relevant expert community, and the general public, remain in ignorance; the warning is lost from view.

Of course, if all ambiguous evidence of harm is published, then the public might be unduly alarmed. This is a very familiar situation in medical research

---

Burden of proof in statistical tests

The conduct of science is influenced right at its core by the choices that are made in relation to methods. The most basic are those used in statistical tests. These convert a collection of data-points into information about correlations between variables. If you look at the results of a standard significance test, you will see something about 'confidence level'.

Although this is a technical term, it means what it says: the sort of confidence that can be placed on the result, that it is free of error. Every test is vulnerable to two sorts of error. We can call them excess sensitivity and excess selectivity. In the former case, there is a greater risk of false results being accepted as true; in the latter, there is greater risk of true results being rejected as false. There must always be a choice between the possible errors that might be produced by the test.

This choice is similar to the 'burden of proof' in legal cases: which side is required to prove its case, and to what degree of rigor? In criminal cases within the English tradition, the testing procedure is highly selective, because we wish to avoid the error of convicting the innocent. The burden of proof is on the prosecution to establish the case 'beyond a reasonable doubt'.

In civil cases, the burden of proof is still with the complainant, but the required 'confidence limit' is not so great, being 'the balance of probabilities'.

on diseases, drugs and treatments. There is no easy answer. But it is clear that the choice of burden of proof for policy-relevant research can reflect the balance of the relevant policy interests.

Where there is more democracy, there will be more concern for precaution and safety; and where less, less. In this way, the most technical of methods, at the core of the research activity, are crucial in the shaping and selection of our knowledge and ignorance, and do so as an indirect reflection of politics.

Openness in science

Science must be made public if it is to be of real use for debate in a democratic society. For a long time there had been a twofold system of publication for science. Inventions were published through patents, whereby the inventor got property rights over their creations;

In some exceptional situations, the burden of proof is on an accused person to establish their innocence. This occurs – with the aid of science – when athletes who have tested positive for banned drugs are presumed guilty unless they can prove their innocence. It had always been believed that science itself is immune to the influence of judgments of this sort; but an understanding of statistical practice shows that the difference is not so great after all.

In research, it is common to guard mainly against accepting false correlations; and so the error of rejecting true correlations is less important. Hence tests will be more selective, and therefore less sensitive. The model of the lab is used for what is called 'sound science'. It might seem that this is totally objective and value-free. But in fact that choice embodies its own values and a balance among possible errors.

By contrast, in environmental monitoring, it can be more important to be alert to possible sources of harm, and so it can be appropriate to have tests that are more sensitive than for lab research.

Tests designed for the lab – and hence more readily accepted for publication – might lead to the exclusion of such warning data, so that we remain in ignorance of possible dangers. To demand a single standard for all sorts of science, is itself a choice. It produces its own possibilities of errors, and its own influence on our knowledge and our ignorance. It also has its own implications for the politics of safety, health and the environment. ∎

and scientific results were published in an open literature, available to all for the small price of being cited whenever used. One system served the economy and the other embodied democracy. Now things are much more complicated and involved. For our present purposes, we note that the 'ownership' of a scientific result can be crucial for its use in society. The scientist may be an employee, in which case her or his firm simply owns all the work. Even if the scientist is nominally independent, they may be on a research contract which gives great or even total power to the funder to control how – if at all – the research is published.

As the system gets more complex, it is more open to stratagems and distortion and abuse. In clinical research, in particular, it is important that the bad news about treatments gets published along with the good news. Otherwise, surveys of 'the literature' will have a serious bias, caused by the sins of omission. Statisticians now use a technique of 'meta-analysis' whereby different statistical studies can be pooled together to provide a large sample for analysis. But if that set of studies is biased by the omission of the negative results, the meta-analysis can be very seriously misleading.

We now know that until recently, pharmaceutical firms routinely concealed unfavorable test results and also made life difficult for researchers who insisted on publishing them. After the exposure of these practices in *Science in the Private Interest – Has the Lure of Profits Corrupted Biomedical Research?* by Sheldon Krimsky, and after legal prosecutions in New York City, the leading firms claim to have turned over a new leaf.[2]

Now that we are aware of our previous ignorance of those practices, we are reminded of the dangers to democracy when the nominally free scientific media fall under the control of vested interests. We might have thought it couldn't happen here in science, but on that, as on other issues of democracy and science, we must now think again.

As safety, health and environmental issues become more critical and polarized, ever more serious abuses of the scientific process are perpetrated. The hostility traditionally shown to whistleblowers is now being extended to critics of any sort. This is only natural under post-normal conditions. When an adverse evaluation of a new product can destroy its prospects, the sponsoring institutions will be tempted to try all possible methods to discredit the bad news.

In post-normal terms, even if the systems uncertainties are low the decision stakes are high, and so the debate will not be conducted within the traditional rules. Scientists who engage in advisory work now find themselves very seriously exposed. Government departments can be even more ruthless than private corporations in their threats to victimize those who publish criticisms. In the UK recently, the members of an official committee were threatened by the Government with lawsuits for defamation if they published a critical Minority Report concerning nuclear plant risks.[3] Under such circumstances, now increasingly common, the integrity of scientists is tested to the limit.

Assessing technology

So far in this chapter we have considered democracy in relation to the sorts of science that are oriented around research on the one hand and around policy processes on the other. But there is a third sort of science, one that raises urgent issues about democracy: the science that is employed in technology, serving industry, commerce and the state. For a long time this sort of science was accepted as essentially beneficial. Even when Karl Marx exposed the cruelties of the Victorian factory system, which was dependent on science and inventions, he accepted it as a progressive stage in the

Big pharma

Biotechnology companies are forming alliances with the pharmaceutical giants hoping to 'bottle money', as Fortune magazine put it. Gene therapy is the holy grail, with companies pinning their hopes on astronomical future profits.

- The University of Chicago estimated the value of reducing deaths from cancer in the US at $46.5 trillion.
- In 2000, 9 of the top 10 biotech companies manufactured medicines.
- In the US, 70% of biotech medicines were approved in the last six years. Four biotech drugs bring in $1 billion annually.
- In 1999, plant-based drugs had global sales of over $40 billion a year. The rewards for patenting plant material are evident.

evolution towards a better society.

Now we know that really bad things as well as good can come from technology, we face the very difficult task of bringing technology under some sort of democratic control. Otherwise, given the power of GRAINN technology to shape our physical environment, our social lives and even our thinking, our formal structures of democracy may become a meaningless sham.

Make no mistake; the extension of democracy in any sphere is a serious business. More democracy challenges the power, the profits and the privileges of those who are benefiting from its deficits. They do not take kindly to such challenges, and they may well believe quite sincerely that their being on top is the best arrangement for all concerned.

When science was in its 'little' phase, it could be argued that on balance it promoted democracy. But mega science, requiring such large-scale financial support, will in many ways be controlled by those with the finance. The selection and shaping of our technologies, along with our knowledge and ignorance, will reflect those constraints.

The first step towards democracy in the control of technology is to be freed of two illusions. The first is that there is an inevitability about technological progress. Just as the facts of science, once discovered, are independent of the circumstances of their discovery, so our technological systems are said to be just what they have to be. Fortunately, we have had enough experience of missed alternatives and failures to know that this is not so.

A high-profile technological failure was the Anglo-French supersonic airliner Concorde. This was a marvel of design and engineering. But it was always marginal as a commercial proposition, being small and also very expensive to run; its prospects were ruined when the first cheap long-distance jet airliners were introduced. Worse, its environmental intrusiveness, mainly the 'sonic boom' which was both unpleasant and potentially costly in claims for damages, kept it off overland routes and hence out of the American coast-to-coast market. This meant that Concorde could never sell enough copies to be viable as an investment.

We are now well on the way to recovery from the second illusion, that ordinary people cannot influence the course of technological development. Concorde is an important example here. Two small pressure groups, in England and the US, kept the issue before the public and the politicians. They prevented the suppression of the critical arguments on Concorde. This made it easier for the airlines to scrutinize the project on its merits; and eventually they backed away. In the event, a few planes were given free to the two national airlines, and they flew across the Atlantic until old age and accidents caught up with them.

Clearly, an awareness of M&M – malevolence and

muddle – and a cultivation of the SHEE – safety, health, environment, plus ethics – sciences will go far to help in this extension of democracy in science and technology. To assist in the development of the appropriate critical skills, I can offer a set of simple questions. These can be considered as a sort of antidote to the traditional mindset expressed in the question 'why not?'. In the case of a proposed new technology, we might ask:

- who needs it?
- who will benefit from it?
- who will pay its costs?
- what happens when it goes wrong?
- who will regulate it, how, and on whose behalf?

Let's not pretend that these simple questions have simple answers. In fact, every new technology is a foray into the unknown. Even the 'needs' cannot really be defined. For once a device comes into use, it can become a necessity; mobile phones are a good recent example.

We might think of the questions as being of a post-normal sort. They are designed to tease out the uncertainties and the value commitments that affect the development and promotion of the technology. They are intended to open a genuine dialogue, in which all sides argue their cases in good faith, ready to learn from the perspectives of the others.

Timing is all important in the societal assessment of technology. As we discussed previously, when a technology is still embryonic it is impossible to know its effects with any accuracy. But if assessment is delayed until all the facts are in, there will be huge decision stakes to contend with, as big investments have been made. The solution is to have different sorts of forums at the different stages of development, and to keep the issues under constant review.

'Upstream engagement' is a good slogan for public involvement sufficiently soon. It offers the only hope of bringing some sort of democratic accountability to the GRAINN technologies.[4] It is already happening with nanotechnology, as the Action Group on Erosion, Technology and Concentration (ETC Group) and others enter dialogue with national and international scientific societies and regulatory agencies (see **Contacts**).

There is, however, another side to those apparently simple questions that can guide public scrutiny. In addition to opening the way to democratic debate, they also remind us of the complexity of our situation and of the impossibility of finding simple solutions of a scientific sort to the urgent problems of science policy. In some cases we might even ask whether a newly discovered 'need' is only an artificially created demand, and in fact a form of addiction. Thus the critical questions also highlight the scientific perplexities that characterize our time.

Poorer countries, science and democracy

All these possible deformations of science are more likely to occur in the societies that we call 'less-developed'. They can become so severe that science, instead of contributing to the development of greater democracy, can actually be used to inhibit it. For in such countries, the national base for science and technology tends to be weak in every way. It does not command great resources of expertise, authority or funding.

External sources for science-related policies, either international governmental agencies or multinational corporations, can be far stronger than local institutions, to say nothing of having enormously greater resources at their command. The consequences are clear, even in the absence of any conscious intent to monopolize science policy. The issues will be defined by those external players, the science will be

Feed the people, nourish the land

A vitally important example of the deformation of Majority – or 'Third' – World science is provided by Colin Tudge in his book on science-based agriculture.[1]

There has been a successful craft of agriculture which, with the aid of science, can feed all the people and nourish the land. This is being destroyed in the interest of total control of food and fiber by multinationals.

From the 1960s on 'Third World' agriculture was deformed by the imposition of a model of agriculture which assumes extensive land, expensive labor, intensive capital, developed scientific infrastructure and predictable climate (as in the North) onto lands where just the opposite holds. That was the well-intentioned delusion of 'aid', which concentrated on the cash-crop grains and ignored pulses, the mainstay of the diet in many countries. It also ignored the native draft animals, and provided no scientific assistance towards their improvement. All that corresponded to 'big science' at home.

Today, with 'mega science' applied to agriculture we have in addition the use of genetic engineering to produce uniform products for the world market, and where the basic science is increasingly oriented to, and controlled by, the private sector. This approach not only destroys the society of the countryside, driving many millions off the land into the city slums. It also destroys the countryside itself. After years of being mined of their water for irrigation, the aquifers are drying up, the water table is falling rapidly, and large areas of the affected continents are becoming desertified.[2]

The solution to the global agricultural water shortage lies in the 'half-forgotten' technologies of local water catchment, rather than more dams or tube-wells.[3] But this would require very local initiatives by craftsmen farmers and hence would be inconsistent with the ruling high-tech agricultural paradigm. ∎

1 Colin Tudge, *So Shall We Reap*, Allen Lane, Penguin, 2004. **2** Fred Pearce, 'Asian farmers suck the continent dry', *New Scientist*, 28 August 2004. **3** (Anon) 'To feed itself Africa must capture more rainwater', *New Scientist*, 28 August 2004.

conducted along the lines they favor, and the results will be owned by them. Any attempt at an indigenous science base, either research or policy-related, will need to play the game in the internationals' way, or else be considered provincial, mediocre, and generally unworthy of support.

Of course, the picture on the ground is enormously varied, just as is the condition of being 'poorer'. But among all the other democratic deficits by which those countries are afflicted, that of science should not be overlooked. Under present conditions, any democratic accountability of research science in less developed countries is largely exercised through the political institutions of the rich countries abroad.

For an example of how this affects the main productive sector, see the box *Feed the people, nourish the land*.

1 Steven J Gould, *The Mismeasure of Man* (Norton, 1981). **2** Sheldon Krimsky, *Science in the Private Interest: Has the Lure of Profits Corrupted Biomedical Research?* (Rowman & Littlefield, 2003). See also the website http://www.tufts.edu/~skrimsky/corrupted-science.htm. **3** Mark Gould and Jonathan Leake, 'Government gags experts over nuclear plant risks', *The Sunday Times*, 1 August 2004. **4** James Wilsdon and Rebecca Willis, *See-through Science – Why public engagement needs to move upstream* (London, Demos, 2004).

8 People's science

Things are changing. Attempts are being made to extend public involvement in policy making about science. Could the result be the creation of an alternative science, complementary to the mainstream?

THE IMPERFECTIONS IN the relationship of science to democracy are, in some parts of the world, well recognized now. It is not so clear how they are to be remedied.

Science is no longer seen as an independent force for material and moral progress, liberating people from material poverty and political oppression. Rather, it is increasingly perceived to be an integral part of the apparatus of business and the business-oriented state.

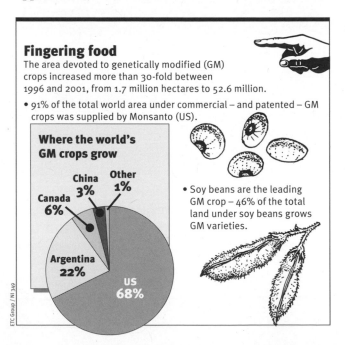

Fingering food

The area devoted to genetically modified (GM) crops increased more than 30-fold between 1996 and 2001, from 1.7 million hectares to 52.6 million.

- 91% of the total world area under commercial – and patented – GM crops was supplied by Monsanto (US).

Where the world's GM crops grow

China 3% Other 1%
Canada 6%
Argentina 22%
US 68%

- Soy beans are the leading GM crop – 46% of the total land under soy beans grows GM varieties.

ETC Group / NI 349

All the criticisms of remoteness and lack of democratic accountability that are leveled against the State in general, are also directed at science in the service of the State and business.

In some respects, this negative attitude is unfair – or at least is becoming out of date. In many areas, and on many issues, the State knows that the old paradigm of decision-making for science and technology policy does not work. It is no longer possible to have 'decide-announce-defend' as a successful way to interact with the public. A great variety of exercises in two-way communication and consultation are regularly mounted by governments on current issues. These are admirable and they could eventually lead to substantial changes in the way that science-related policies are framed and decided. This is part of a general change in the mode of governance in modern societies, where 'representation' through occasional elections is complemented by 'participation', involving regular consultation with the NGO sector and special forums.

These efforts suffer from two major weaknesses. The first is that, in general, the State knows in advance of the consultation what it wants, which is usually what business wants. After all, growth needs technology and if we veto new technology we will inhibit growth. And this will not happen uniformly all over. If some nation or group of nations takes a negative attitude to particular innovations, they will simply move elsewhere and we will be left behind. From this perspective, those who oppose the innovations are then seen as misguided and perhaps deeply malevolent as well. Do they really want us to be left behind in the race for new technologies? For even if winners don't always take all in the new technologies, you can be sure that losers will lose all.

The opponents of new technologies frame their arguments in a broader perspective. They cite the risks of unknown and unknowable future contingencies

– 'ignorance' in other words. They might also bring in considerations of lifestyle and about what sort of world we really want to live in. Their visions suffer from being obviously vague and utopian, while those of the proponents of growth can at least appear to be precise and realistic. So the two sides argue past each other, the government/business side for GRAINN and the opposition for SHEE (using M&M for evidence). And when the side with the power is organizing the debate, the other side will naturally distrust the way it goes. For them, this is not democracy or participation, but an exercise in persuasion or in neutralization of the opposition.

The other weakness of these methods of extension of democracy is that they are inevitably of a top-down character, or nearly so. The machinery of consultation is, even with the best will in the world, rather artificial. Ordinary people do not normally attend focus groups to discuss the issues of the day, any more than they normally engage on jury duty in the courts. The experience of those new forums can be very educational for participants and for those who have organized their discussions; but it is still something specially constructed. Participants are still left wondering whether they have had any real influence on policy; and if they did, by what right did they do so? They run the risk of being seen either as impotent or as illegitimate. There is no easy way out. But a recognition of these contradictions in participation can prepare us for creative solutions. See 'The Ladder of Participation' by Sherry Arnstein.[1]

Participatory science: what could it be?
Creativity in governance will not be achieved only by deliberately planned top-down policies, important as they are in their own way. They must be complemented by all sorts of innovations. At first these will seem untidy, perhaps even unruly, and lacking in

coordination and quality control.

Such innovations have been occurring in local affairs and social policy for some decades now, as groups of citizens get together not merely to protest but also to resolve the issues that are damaging to their lives. Indeed, the sophisticated modern state now welcomes such initiatives at community level and fosters them, sometimes even co-opting or adopting those that are successful.

Similar creative processes for science have come into being recently. Perhaps it took longer for citizens to overcome the mystique of science and of professional expertise; but that process is well underway now. Innovations in policy-related science occur in all sorts of areas and with all sorts of relationships with authority. Their very diversity is a sign of their vitality and of their importance for the future of society and of science itself.

Of course, the 'science' that is involved is only rarely that of the research laboratory. Indeed, along with the 'extended peer community' of post-normal science, who have their 'extended facts' – deriving from their own experience and from irregular or unorthodox sources – they are also creating an 'extended' idea of science. This is not restricted to the established knowledge as created and certified by the recognized institutions; rather it is knowledge that can be used in problem-solving and struggle by people wherever they are, and whatever the issues they confront.

The biggest hurdle to overcome is the prejudice that no-one can deal with science unless they have a university degree in the special subject. This is well ingrained among scientists, even those that really want to make their subject more democratic. All too often, well-intentioned university courses in 'science for the citizen' get so much packed into the syllabus that it would be a rare scientist who could pass the exam!

But we already know that people can make effective

use of science while having a less advanced training; all sorts of professionals, practitioners and technicians in engineering and medicine do so regularly. They need to know enough science to explain what they do with their tools and – even more important – to understand what is happening when those tools don't work. And experience has shown that citizens can do the same, in their way, for their needs. They have been entrusted to do so on juries in cases involving science and now they are showing their competence in consultations on policy issues. Indeed, one of the great lessons of all the consultation exercises is that it is frequently easier for citizens to understand the relevant science and the scientists' point of view, than it is for the scientists to grasp the citizens' special concerns.

There is always the risk that citizens will be misled by clever manipulators, in these issues as in so many others. Or they will find it impossible to admit either that the science is inconclusive, or worse, that the science supports the other side. In the context of an embattled campaign, both sides will use and perhaps twist the facts to their own benefit. These dangers are real. But no-one can now plausibly argue that industry, government, or even an entrenched scientific expert group, are totally immune to such perils. Dirty tricks can be found on all sides. Rather like in a jury trial, the 'court of public opinion' will, with all its obvious imperfections, be the best protection for fair play.

It is in such conditions that science and democracy, and in particular a free press, are mutually supporting. For if the dominant side can conceal facts and victimize critical scientists, then disasters will inevitably occur. This is what happened in the BSE debacle in Britain in the 1980s to 1990s, during which contaminated beef was declared safe. Since there is a relatively free press in the country, the corruption of the authorities was eventually exposed for all to see.

Some scientists might legitimately wonder whether

high-quality work could possibly be done in these new forums of post-normal science. However worthy and insightful these 'extended peers' might be, by definition they don't know the science! There are three aspects of

Lessons from the sick school

All the elements of a film script, a worthy sequel to *Erin Brockovich*, are present in the story of the discovery of pollutants in a new intermediate school in Ohio. The story shows how post-normal science works in action. Parents, teachers and children were delighted at first when their much needed new school was built. But everyone at the school was soon affected with unpleasant symptoms, including burning eyes and nasal passages, sore throats, nausea and fatigue. As if to confirm that something was wrong, mold began to appear everywhere. The building representative for the school was herself a science teacher; but her training had given her no preparation for this situation. Eventually she learned a relevant 'extended fact', when a nurse told her of 'sick school syndrome'. She informed the principal and school superintendent. Appreciating the very high decision stakes in the costs of repairs, the latter decided to sit on the problem. Some months later he presented the teacher with an official scientific report stating that key pollutants were below detectable levels. (This was later discovered to be inaccurate).

Everyone continued to suffer in a general ignorance, until some six months after the opening of the school, when an 'extended peer community' got on the job. A local TV station told the story of a 'mysterious illness' at the school. On the basis of this unorthodox scientific support, an action group was formed. But they still got no positive response from the authorities. By this time the teacher was so ill that she had bronchial spasms and needed an inhaler. Eventually in the springtime there was a demonstration in the form of a 'teach-out', classes being held outside the school. As the teachers were being ordered back into their classrooms, many parents arrived to remove their children. The resulting publicity was sufficient to ensure the closure of the school. The struggles did not end then; but the story of the polluted school became known all over the state, as an example of what can be done. The 'normal science' of the teachers' syllabus and of the official experts did not help these real people with their real problem; but in the end they became competent to recognize their situation and to do something about it. ■

Paul Rather, 'Parents and Teachers Shut Down Moldy New School', *Everyone's Back Yard*, 20/4, Winter 2002.

these new forums that enable us to be confident about the possibility of their achieving high quality in the science as well in their political functions.

First, while technicalities may well be at the core of the inquiry, they are not the whole. We recall from the road safety example that the policy issue itself must first be settled – and then the research priorities and problems defined. These are done by people in particular positions, such as politicians, civil-servants, representatives of NGOs. They will have their own agendas and perspectives. These can be scrutinized and criticized; if they are not, these decisions may well go by default. Similarly, at the end of the process, there will be decisions about the ownership and publication of the findings.

Who takes these decisions, by what criteria and procedures – and who appoints them – are also crucial questions. On all these, non-scientists are fully competent to see what is going on and to ensure balance. Otherwise, the knowledge that eventually emerges (as well as the ignorance that is imposed) may well be selected and shaped in ways that do not reflect a full and fair perspective on the issue.

At the technical core of the inquiry, the content must relate closely to the real world of practice and experience. Even if some very abstruse science is involved, what counts is how well this describes some aspect of an issue of concern. In case of need, there can be a 'counter-expertise', either specialists within an interest group, or experts hired for the occasion. Needless to say, there are very serious problems in translating between the experts and their constituents. There are also dangers that experts will be absorbed into the game of trading technicalities with the insiders. They too are stakeholders, with their own special perspective and agenda. But with full awareness of such pitfalls, the experts can do their job.

Finally, there needs to be some shared commitment to

quality, analogous to the morale and idealism that are required for the prevention of corruption in research science. This element can be described as 'negotiation in good faith', which we have already mentioned. This concept is well established in many proceedings worldwide. It is sufficiently clear in practice, for legal sanctions can be applied when one side fails to respect it. Technically trained experts are no better equipped to practice this than are citizens. With such a regulative principle in place, there is no reason why dialogues in post-normal science situations should be lacking in the means to assure quality.

Science shops movement

Some sorts of participatory science are officially promoted, others arise out of immediate need or struggle, and still others relate to broad movements of citizens' and consumers' lifestyle commitments.

There is now an international 'science shops' movement. This got its start with 1960s radicalism in The Netherlands, then became established in the university system there. Eventually it got support from the European Commission and is now branching out all over the world.

Its principle is to provide information and advice to people and not-for-profit organizations, on questions with a scientific content. Sometimes the question can be answered from existing knowledge; sometimes it calls for an information search, or a special research project. Where there is a link to a university, students or even staff can get research-time credit for the inquiry. The variety is enormous; each science shop creates its own clientele and style. Sample projects include: tackling isolation of senior citizens, monitoring of noise from wind turbines at night, environmental education in junior schools, micro-finance for Majority-World housing, and land-use mapping and planning in a heavily industrialized region.

People's science

It might appear that such projects are of the sort where the experts, however enlightened, still dole out facts to needy citizens. But that is to miss the basic point that the citizens are the clients who come to the science shop with a perceived issue. And the service starts with a dialogue on that issue, clarifying it and then seeing whether it can be resolved by the disciplined study of a scientific problem. The trained scientist brings one sort of expertise to the dialogue, and the citizens bring their problems and their perspectives. Their motto is 'living knowledge'. For them, scientific knowledge truly lives only by being transformed so as to empower ordinary people in their daily lives.

The science shops movement is now broadening its scope to include community-based research. This movement, existing mainly in the US, has grown up in parallel with the science shops movements and is now supported by some of the most prestigious universities.[1] In an exciting new development, there is now a Worldwide Virtual Network of Young Practitioners Working on Science and Society.[2]

Other movements, focusing somewhat more on the techniques than on the participatory elements, are Alternative Technology – founded by the visionary EF Schumacher – and Permaculture.

There is no sharp dividing line between such non-political activities and those arising out of struggles by communities for their health and safety. In the latter area, one of the senior organizations in the US arose out of the discovery of pollution in the community called 'Love Canal' near Niagara Falls in New York State.

Now operating as the Center for Health, Environment and Justice, its focus is on community activism, particularly – but not at all exclusively – among people of color and others who are exposed to industry-based pollution. They provide excellent guidance on the strengths, limits and pitfalls of using expertise, both

Corn syrup's obese credentials

Corn syrup – made from what Europeans call maize – is a key element in the tangled tale of the obesity epidemic that started in America and is now a worldwide scourge.

This product has many advantages as a sweetener, and has been widely used in all sorts of processed foods ever since the 1970s. Even back then, some scientists realized that its main component, fructose, behaves differently in the body from other sugars such as glucose. In particular, it is not broken down into constituents but goes straight to the liver to be metabolized there. After several decades of steadily increasing use of corn syrup, researchers began to take it seriously. They discovered that test animals fed excessive fructose developed resistance to insulin. Thus, a diet heavy on fructose not merely accelerates obesity, it also contributes to diabetes. This contradicted the ruling paradigm of nutritional science, which had focused on fat rather than sugar as the root of all evil. It was also bad news for the junk food industry. Here we have an illuminating interaction between commercially driven lifestyle, industry, science, and ignorance of ignorance, leading eventually to a large-scale socio-medical problem, to which the authorities have recently made a vigorous, if belated, response. ∎

Greg Critser, *Fat Land* (Houghton Mifflin and Penguin 2003).

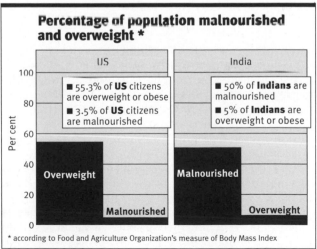

Percentage of population malnourished and overweight *

US
- 55.3% of **US** citizens are overweight or obese
- 3.5% of **US** citizens are malnourished

India
- 50% of **Indians** are malnourished
- 5% of **Indians** are overweight or obese

Per cent (0–100)

US: Overweight / Malnourished
India: Malnourished / Overweight

* according to Food and Agriculture Organization's measure of Body Mass Index

FAO / NI 353

legal and scientific, in community struggles. The great lesson is that neither lawyers nor scientists will win a campaign; only the determination, skill and especially the integrity of the campaigners and the community they represent can accomplish that.[3]

An important bridge between consumerism, militancy, lifestyle and contested science is food. In Europe a new focus of politics lies in the battles between the pressure groups and the governmental-industrial-scientific complex over genetically manipulated foods and crops, with the supermarkets playing it safe in the middle.

Defying official scientific advice, consumers reject GM and increasingly turn to organic produce. The 'slow food' movement in Europe gathers strength. The official side is weakened by the immature and confused state of its core science, nutrition. People have discovered that the current mass epidemic of obesity – with a resulting increase of diabetes – has been caused largely by high-technology junk food. The official nutritionists' advice had mainly been irrelevant or wrong. In this area at least science has lost its mystique.

The full range of styles of activism and science is now present and flourishing in patients' movements and organizations in medicine. In some cases, a particular disease will have a more 'tame' group relating to pharmaceuticals and research, along with a patients' group devoted to questions of treatment and care, and finally a militant group that raises the uncomfortable issues that nobody else wants to know. In medicine, the internet has already transformed some aspects of practice. Patients can connect to sources of information, and to each other, and thereby build and deploy a 'living knowledge'. This enables them to enter dialogue with the doctors, not as exact equals but as possessors of a complementary knowledge that should and does command full respect. With such patients as an extended peer community, medicine is itself entering the post-normal age.

Medicine could also provide the societal basis for what we might call a future 'complementary and alternative science'. The medicine with that title represents a quiet mass revolt, already large and still growing, against official science. Mainstream medicine, based on conventional science, still operates mainly on the assumption that the human body is a collection of bags of physiological processes suspended on a bony frame.

Disease in the living person is to be analyzed and treated as if it is no different from reactions in laboratory glassware. Of course this approach is very powerful and up to now has been largely beneficial. But it no longer commands implicit trust. By contrast, research in Complementary and Alternative medicine is fragmented, underfunded and still searching for an appropriate methodology. But the need and the opportunity are there, and the science will mature in response.

The consumers' mass migration to a different sort of medicine is not merely a choice between treatments. It involves issues of lifestyles, and indeed of reality. Whether it be called 'integrated', 'holistic', 'complementary' or 'alternative', this medicine involves a vision of the person as essentially more than a machine with a ghost inside. Whether using Eastern concepts of *chi* or *prana* or the Western idea of 'subtle energies', it is based on a reality that was officially denied in the 17th century and has been suppressed and derided ever since.

As complementary and alternative medicine struggles on the great variety of fronts against entrenched opposition, it becomes involved in a new politics of reality. This is not refracted through religious doctrines, as it was in the battles of the Protestant Reformation in Europe. Now it concerns the way people see, feel and conceive themselves in the world, extending from before birth to after death.

The huge variety of treatments in this complementary

and alternative tendency, all of which coexist largely happily, is a reminder that the relevant conception of knowledge is not one of the 'single correct answer' as in the science textbooks. Rather, the therapeutic practice is a creative, participatory dialogue of mutual learning between healer and client. Any science that is genuinely related to this movement will move away from the alienated style that conventional natural science uses to ensure 'objectivity'.

Appropriate modes of achieving and ensuring rigor can be adapted from existing practice in the humanities and social sciences. There, the presence of a reflective, conscious investigator is not denied, but is built into the understanding and management of the work.

It is impossible just now to predict what sorts of arrangements and accommodations there will be, between a new complementary and alternative science and the mainstream. We could imagine two sorts of tendency. The 'complementary' wing of medicine is what we have been calling 'post-normal'. It would grow out of the SHEE sciences, with an enriched societal base as in the science shops and community research. The 'alternative' sciences – in this sense – would conduct studies into the enhanced consciousness and subtle energies that are at the core of the new therapies. And, in the spirit of these new times, there would be a fruitful collaboration and sharing of experiences and perspectives.

As yet, there is no connection at the grass roots between these two embryonic tendencies. There are many who would accept a 'complementary' science as an extension of democracy, while still vehemently rejecting the philosophically radical 'alternative' science as superstition and fraud. But until quite recently the whole of complementary and alternative medicine was rejected in just that way. Now it is largely accepted by mainstream medicine, even if only on pragmatic grounds. But one could imagine future mass campaigns on lifestyle politics in which consciousness is raised and transformed and out of which could come a synthesis of the two streams.

The last half-century has seen the creation of a powerful new consciousness on a great range of issues including: world poverty, civil rights, nonviolence, the environment, feminism, personal experience and growth, and multiculturalism. In addition we have respect for those with disabilities and with different sexual orientations, as well as for non-human sentient beings.

In retrospect the list is very impressive, even more so if we recall how, certainly up to World War One, prejudice of every sort, based on color, class, nationality – to name a few – was rampant and respectable. This sort of progress in civilization shows how the future can really be different from and better than the past. Similar future innovations in relation to the big themes of science, politics and reality are not inconceivable. When, where, how and indeed whether this might occur, is a matter for the future.

We in the rich countries know too little about what is happening on the ground in the poorer nations, where the two tendencies sometimes coalesce. As we have already imported so many ideas from their cultures about healing and being, this new synthesis might well eventually be another of their gifts to us. The internet will be enormously important in enabling the rapid and inexpensive exchange of information. Some social movements, such as Appropriate Technology, promote ingenious, simple techniques along with traditional medicines and social reform.[4]

An inspiring set of examples of what can be done in every social milieu, rich and poor, is in the monograph by Tom Wakeford, 'Democratising technology / Reclaiming science for sustainable development', on the website www.itdg.org/docs/advocacy/democratising_technology_itdg.pdf.[5] An important example of how 'we' can learn from 'them' even as we try to help, is the work on ecosystem health by David Waltner-Toews and his colleagues; see the website http://www.nesh.ca/

All this agitation might seem to be a very far cry from 'science' as we have grown to understand it. We might well ask, where are the 'scientists' among all those activists and healers? But it is useful to remember that there was, not so long ago, a time when one did not need to have a PhD in order to contribute to the growth of knowledge and technique.

Through the 18th and 19th centuries, in Europe and America, it was possible for a person to educate themselves up to the forefront of knowledge in a practical or even theoretical field. The very idea of a 'scientist' did not exist until it was announced in the 1830s, and even then it described a calling rather than a job.

In retrospect, we can see 20th century science as a temporary period of dominance of professional research practice, just as 20th century medicine was a temporary period of dominance of scientific medicine and the 'germ theory' of disease.

The 'extended peer community' of post-normal science is only now finding its identity. Going further afield, we should recall that there was a great stretch of centuries during which the various civilizations of the East had levels of culture, including science, compared to which Europe was barbarian. And now once again, the future is open.

1 Arnstein, SR (1969) 'A Ladder of Citizen Participation,' JAIP, Vol. 35, No. 4, July 1969, pp. 216-224. See http://lithgow-schmidt.dk/sherry-arnstein/ladder-of-citizen-participation.html **2** www.scienceshops.org and www.loka.org **3** http://alba.jrc.it/science-society/ **4** www.chej.org **5** www.changemakers.net

9 Science, its future and you

The energy of science flows from one world center to another. And the vigor of science depends on the commitment of young people considering it as a vocation. The old assumptions of the perfection of science are obsolete. In their place we have questions.

WE HAVE SEEN how, in spite of the apparent timelessness and objectivity of science presented in textbooks, science is both diverse and ever-changing. What will come after mega science? Will a 'complementary and alternative science by the people' ever make any real difference, and if so, how long would that take to happen?

Acknowledging our ignorance is one of the main lessons of the post-normal approach – and let us do so here. The message here is not predicting the future, but rather opening our imagination to what might be starting to happen.

There is another lesson from history, about how the creative energy in science comes and goes. This phenomenon is well known in the artistic fields, but it is surprising to find it in science. Overall, there has been a remarkably steady increase in the quantitative indicators of scientific activity since the 17th century, with a doubling time of 15 years. But within that trend, there has been a succession of national centers of excellence, each one taking the lead for a few generations, and then giving way to others.

Going back to the early 17th century, we find Italy in the lead, but then England took over, with the generation of Boyle, Hooke and Newton. Scotland – with provincial England – made a strong showing in the 18th century. France produced a generation of genius centering on the period of the Revolution. As they lost their élan, the Germans became dominant, with their researchers creating the complex social apparatus of professionalized research. Right up to World War Two, any serious student of chemistry had to know enough German to read the scientific literature in that language.

Most recently, the Americans, first helped by the refugee scholars and then by the munificence of the military, established a commanding lead that seemed unshakeable. But there are signs of an aging of the corps of scientists there as well as in Europe. American science depends increasingly on Asian migrants, just as China and India are now finding their way to scientific strength.

No-one knows how this cyclical effect occurs; but up to now it has seemed remarkably regular. And with each cycle, the dominant style of the social activity of science changes. This is why we need not accept that the contemporary mega science will rule forever, even if we cannot predict what will supplant it.

There are now encouraging signs that the research science community is starting to fight back against the excesses of mega science. There is a strong movement for scientific papers to be freely accessible on the internet. At present someone who wants to consult a paper must either have a subscription to the journal – which can be very expensive – or have direct access to a science library, or be prepared to pay a substantial sum for a copy. It is easy to see how this situation reinforces the advantages of those who already have a privileged position in the competitive world of research.

The reform movement takes its name from the 'open source' movement in computer software, that is now breaking the monopoly control of key programs by making the basic 'source code' accessible to all. In the case of science publication, the most modest proposals call for papers to be available after a certain interval, rather like patents.

The more radical proposals call for the whole cycle

of review and revision to be public. Then anyone with a concern and the relevant competence could follow the criticisms and replies, and form their own judgments on the quality of the work. There are enormous practical difficulties; for example, many learned societies depend on the income from publications to balance their books. But once the idea is there, some sort of reform is inevitable. This issue might be the one that forms the nucleus of a movement for a general reform against the self-destructive tendencies of mega science. The struggles over that reform might provide a challenge to research scientists who see beyond the confines of the lab.[1] This development could be vital, for high-quality science cannot long survive without a new generation of committed recruits.

There are many factors that induce talented young people to choose a particular career. We simply do not know what happens to those wellsprings of commitment, that for a time inspire talented young people and then fade. This does seem to be happening right now, all over the Western world. But we can discover something about ourselves and see whether potential recruits still find science to be a worthy challenge for committing their life's endeavor.

To that end, of self-discovery, I have produced this questionnaire for students who know something about science. It must be confessed that this document is radical. It raises questions about science that have hitherto been totally absent from teaching within science or even about science.

Although it contains implicit criticisms, it is not intended to turn people off science. Rather, it is designed to help them get a better idea of why they might want to engage in science as a career or as a vocation, either within mega science or in some other way.

There are two levels of questions to be posed. The first set are more obvious; they concern science in society and how it is to be made more accountable.

Science, its future and you

They are useful as introductions but they do not penetrate into the core of the problem, the conduct of science itself. This second set are more personal. Their primary focus is students, for they know enough about science to feel confident in questioning it, but are not yet so committed to a career that they are inhibited from criticism. Of course, anyone, not only students, can engage with these questions as a way of increasing their own awareness and understanding.

Before stating the questions, I want to remind us of the sort of conceptual box in which we have been placed by generations of propaganda for science. This is described by a list of assumptions about science, which until recently were sheer unquestionable common sense. For anyone living inside that set of assumptions, questions like the ones in the subsequent lists are perverse. But for those for whom these old assumptions are obsolete, the new questions are an urgent necessity.

The old assumptions

- Science is coherent, objective, unproblematic and well-bounded.
- Science is central to decisions about practical action in everyday life.
- Science is unencumbered by social and institutional commitments.
- Uptake of science is determined by intellectual ability.
- Ignorance on the part of the public has to be remedied.
- Unscientific behavior results from the failure to apply scientific knowledge.
- Scientific thought is the yardstick with which to measure the validity of everyday thinking

These assumptions might be thought of as a sort of

catechism, articles of faith which both define a belief system and also guide practice. We notice that they are all *about* science, not scientific statements themselves. Yet those who subscribed to them believed that they are as obviously true as the atomic weights of the chemical elements.[2]

By contrast we now move to:

The political questions

- Who decides on the priorities and resources, whereby we have the possibility of knowledge in some domains and remain in enforced ignorance in others?
- Who decides on the ethics of research, including the creation of possibilities for harmful technologies, and the infliction of pain on sentient beings of all sorts?
- Who should decide – and by what sorts of arguments – on those applications of science that can alter our constitutions as human beings?
- Who assesses the consequences, intended and unintended, of scientific and technical advance; and how is democratic accountability in science to be achieved?

And then on to:

The personal questions

- How can I engage with science to make a better world?
- How can we best deploy science to prevent further harm to the biosphere, that we see in climate change and species loss?
- How can we stop the use of science in biopiracy and other forms of exploitation of Majority-World people?

Science, its future and you

- How can official scientists regain trust for the next occasion that they reassure the public that something is 'safe'?
- How can science prove that something is impossible, like acupuncture or homeopathy?
- How can we rescue textbook science from its implicit message that for every problem there is just one correct answer?
- How could education in science help us to teach ourselves and to criticize what we are taught?
- How can citizens become skilled in the assessment of the quality of policy-related scientific information?
- How can science be taken out of the lab into the community, solving people's real problems in rich and poor countries alike?
- What can science students do to make science education more relevant to the real world?
- If I get a job as a scientist, who will direct my research and who will own what I discover?
- How can science help me to build my life?

These questions come at the end of this book, not at the beginning. This book is not a textbook, providing answers. Rather, its purpose is to be an introduction to the questions. The reader is invited to take them as invitation to explore science for themselves.

Over to you!

1 David Bollier, Leveraging Scientific Commons to Foster Innovation, The Networker Vol. 11 #1, http://www.sehn.org. See also his website http://www.bollier.org
2 D Layton *et al, Inarticulate Science? Perspectives on the Public Understanding of Science and Some Implications for Science Education* Driffield, E Yorks, Studies in Science Education, 1993.

Contacts

The issues raised in this book are being discussed increasingly in all the national and international forums for science. Here are some organizations that have special concerns.

Appropriate Technology
website: www.changemakers.net
email: cmbcbi@ashoka.org
tel: +91 33 2483 8031 (India);
+1 703 527 8300 (US)
Promotes ingenious, simple techniques along with traditional medicines and social reform.

Center for Health, Environment and Justice
website: www.chej.org
tel: +1 703 237 2249
Leads grassroots campaigns in the US against pollution and environmental hazards.

ETC (Action Group on Erosion, Technology and Concentration)
website: www.etcgroup.org
tel: +1 613 241 2267
Formerly RAFI, based in Canada, operates worldwide. Currently leading the campaign on nanotechnology.

Friends of the Earth
website: www.foei.org
tel: +31 20 622 1369 (International)

Greenpeace
website: www.greenpeace.org
tel: +31 20 5148 150 (International)
Both the above engage with science-related issues.

RIVM (Dutch Environmental Agency)
website: www.nusap.net
Web-based, for guidance on uncertainty management.

Intermediate Technology Development Group
website: www.itdg.org
tel: +44 1926 634400
Follows the vision of EF Schumacher, promoting practical measures for self-help and development.

Worldwide Virtual Network of Young Practitioners Working on Science and Society Issues
website:
http://alba.jrc.it/science-society/
Web based, post normal science network sponsored by the European Community's Joint Research Centre.

Science Shops
website: www.scienceshops.org
tel: +31 30 253 7363
Netherlands-based science shops movement, linking to similar groups internationally.

Community Based Research
website: www.loka.org
tel: +1 301 583 9398
Allied US movement.

Contacts

International Association for Science, Technology and Society
website: www.nasts.org

Union of Concerned Scientists
website: www.ucsusa.org
tel: +1 617 547 5552
Leading US organization of scientists, making authoritative pronouncements on public and environmental issues.

Science and Engineering Ethics (journal)
website: opragen.co.uk.

The James Martin Institute for Science and Civilization,
Oxford University, website: www.martininstitute.ox.ac.uk.

esocialaction@email-lists.org
A list connecting people engaging in radical social action in all spheres

Science and Development Network
website: www.scidev.net
tel: +44 20 7292 9910
UK-based, works in Africa, Asia and Latin America to promote regionally appropriate science.

Scientists for Global Responsibility
website: www.sgr.org.uk
website for worldwide network: www.inesglobal.com.
tel: +44 7771 883696
UK-based organization that gives lead on many of the issues raised in this book.

The Science and Environmental Health Network
website: www.sehn.org

Bibliography

Bill Bryson, *A Short History of Nearly Everything*, Doubleday 2003
Fritjof Capra, *The Hidden Connections: A Science for Sustainable Living*,
Flamingo/HarperCollins 2003
Steve Fuller, *Science*, University of Minnesota Press 1998
John Gribbin, *Science – A History*, 1543-2001, Penguin 2002
Mayer Hillman, *How We Can Save the Planet*, Penguin 2004
Candace Pert, *Molecules of Emotion*, Pocket Books /Simon & Schuster 1999
Jerome R Ravetz, *Scientific Knowledge and its Social Problems*, Oxford
University Press 1971 (UK), Transaction 1996 (US)
Martin Rees, *Our Final Century*, Heinemann 2003 (UK). Our Final Hour:
*A Scientist's Warning: How Terror, Error, and Environmental Disaster Threaten
Humankind's Future In This Century – On Earth and Beyond*, Basic Books, 2003 (US)
Iwan Rhys Morus and Peter J Bowler, *Making Modern Science –
A Historical Study*, University of Chicago 2005
Zia Sardar, *Introducing Science*, Icon Books (UK),Totem books (US) 2002
Dr Keith Scott Mumby, *Virtual Medicine*, Thorsons 1999
Tom Wakeford, *Democratising technology: Reclaiming science for sustainable
development*, Intermediate Technology Development Group 2004

Other brief citations to authors and publications appear in the text.

Index

Bold page numbers refer to main subjects of boxed text.

accidents
    nuclear 52, 72
    road 101
accountability, scientific 131
acetone 98
Action Group on Erosion, Technology and Concentration *see* ETC
action research 76
acupuncture 35, 132
addiction
    and need 109
affirming the consequent 86
Africa 34
Agent Orange 51
aggression gene 59
agricultural research 6
agriculture, science-based **110**
AI *see* Artificial Intelligence
aid, overseas 110
aircraft, supersonic 6
Alamogordo, Mexico 30
algebra 26
alternative energy 69
alternative technology 120
ammonia 98
anecdotal evidence 93
animal experimentation 8, 60
anti-Nazi scientists 95
Aotearoa/New Zealand 54
applied ('normal') science 18, 70, 71
Appropriate Technology 126, 133
architecture 39
Argentina 112
Aristotle 21
arsenic 98
Artifical Intelligence (AI) 10, 13
    see also GRAINN
asbestos 98
Asian migrants 128
assumptions about science 8, 130-1
astrology 81
astronomy 22-5, 30, 37, 41, 44-5
athletes
    tests for banned drugs 103
atomic bomb 37, 50, 52, 66, 82, 95
atoms 29
    structure **30**, 88
Atoms for Peace program 52
Australia 54

Bacon, Francis 17, 25, 26, 36
Bell labs 53
'best-fit' trend line 64
big science 50-2
bio-engineering 11-12
biology 11, 26, 52, 68
biomedical research 104
biopiracy 8, 55-6, 131
bio-prospecting 8
biosphere, harm to 131
biotechnology
    and pharmaceutical companies **106**
black bodies
    energy emission 29
blood circulation 42
Blundell, Sir Tom 6, 17
Bohr, Nils 43
bombs *see* atomic bomb
bouncing effect 87, 88
Bovine Spongiform Encephalopathy *see* BSE
Boyle, Robert 127
BP Amoco 68
Bradley, James 24
Brazil 126
Britain 53, 54, 105, 118, 127
Brownian motion 29
Bruno, Giordano 41
BSE (Bovine Spongiform Encephalopathy) 6, 93, 118
Burbank, Luther 96
burden of proof **102-3**
butane 98

cadmium 98
campaign organizations
    telephone/website contacts 133-4
Canada 54, 112
cancer 13, 106
carbon concentrations 84
carbon monoxide 98
careers 52, 58, 96
Carson, Rachel 31, 32, 51, 82
Carver, George Washington 96
cash crops 110
Catholic Church 24, 41, 42
    see also Christianity
cell lines
    contamination **57**
Center for Health, Environment and Justice 122, 133
certainty 62
    see also infallibility of science

chemicals 32
    environmental releases 6
    nomenclature system 43
    in tobacco smoke **98**
chemistry 11, 43, 128
Chevron Texaco 68
children
    diseases 75
    susceptibility to environmental hazards 74
China 22, 36, 43, 112, 128
Christianity 22
    see also Catholic Church
citizens' science 76, 116, 132
civic science 76
civil engineering 72
civil rights 125
climate change 6, 15-16, 70, **84**, 91, 131
cloning of humans 32-3
coal lignite 68
Coghlan, Andy 57
Cold War 50
common cold 13
Commoner, Barry 83
community activism 122
community-based research 76, 120, 132, 133
community participation initiatives 115
compasses, magnetic 36
complementarity *see* yin-yang
complementary medicine 123-5
computers 10, 13
    hacking 14
    simulations 53
    see also software
Concerned Scientists, Union of 134
Concorde 107
confidence levels, statistical 102
consequences of science and technology 14, 15, 92, 131
consequences, unintended 14, 15
consultation processes 114-15
continental drift 45
control of science 14, 51, 52, 53, 128
Cook Islands 85
Copernicus, Nicolaus 23, 25, 41
corn syrup **121**
corporations 109, 110
    biopiracy 55-6
    control of science 14, 52
    renewable energy research funding **68**, 69

Index

corruption 32, 49, 56, 59, 92, 104
  prevention of 119
cosmologists 79
cosmology 24
counter-expertise 119
Creationism 28
critical science 76
criticism of science 26, 59, 65, 105, 118, 132
crops 35
  genetically-modified **112**, 122
cyborgs 13

dams 17, 110
Darwin, Charles 27-8, 43
Davy, Humphrey 43
DDT 31, 98
decision-making 77
  questions 131
defoliants 51
democracy
  and science 94-111
DES 93
Descartes, René 25, 81
The Descent of Man 28
desertification 110
diabetes 16, 121, 122
dialogue
  benefits for scientists 77
  patient/doctor 122-3
  on policy issues 74, 75
The Dialogue on the Two Great World Systems 41-2, 62-3
dictatorships 95
disabled persons, respect for 125
disagreement among scientists 78-9, 132
diseases, microbial 16
DNA 52
dogmatism, scientific 79
drugs, medicinal 5
  clinical trials 5
  long-term effects 6
  toxicity 5
Dutch Environmental Agency *see* RIVM
dying, doctrines of 36

early warning systems 92
earth cooling theory 44
Easter 22, 23
Eastern scientific tradition 35-6
Eckstein, Rabbi Josef **75**
ecological science 76

economic growth 94, 113
ecosystem health 126
Edison, Thomas A. 96
Edmondson, K. 5, 6
education
  functions 91
  see also teaching *and* textbooks
egalitarianism 96
Egypt 35, 36
Einstein, Albert 29-30, 44, 45, 66
Eisenhower, President 53
electricity costs, comparative **68**
embryonic stem cells 32
energy systems **68**
environment 30-2
  degradation 37
  effects of released chemicals on 6
  monitoring 103
  see also SHEE
environmental education 76, 120
environmental hazards 74
environmental issues 30-1, 32, 50-1, 125
Erin Brockovich 117
Erosion, Technology and Concentration, Action Group on *see* ETC
errors in science 34-46
  admission of 45
  statistical tests 102
error bars 85, 87, 88
ETC (Action Group on Erosion, Technology and Concentration) 109, 133
ethics 26, 32, 60, 131
eugenics 98
European Commission 119
European Environment Agency 92
evolution 27, 28, 43
experimiental samples
  error sources 86
extended facts 76
extended idea of science 115-16
extended peer community 76, 77, 93, 117, 118
ExxonMobil 68

fairness of procedure 95
feminism 125
Fermi, Enrico 50
Fiji 85
The Final Century 16
fine-structure constant **88**

fishing 6
flood control 17
FOE *see* Friends of the Earth
food production 32
food science 122
Foucault, Jean 25
France 34-5, 127
Franklin, Benjamin 81
freedom of inquiry 95
French Polynesia 85
Friends of the Earth (FOE) 133
fundamentalism 28
funding, research 58, 93
Funtowicz, Silvio O 70
future of science 11, 34, 127-32
  see also predictions

Galapagos 85
Galileo Galilei 23, 26, 39-42, 62, 80
Galston, Arthur 51
gay gene 59
gender-bending pollutants 16, 70
gene therapy 106
genes 10
genetic diseases 75
genetic engineering 110
genetically-modified (GM) crops 8, **112**, 122
genetics 28, 59
genomics 10, 11-12
  see also GRAINN
geological record 27, 43
geometry 21, 26
Germany 95, 127, 128
Global Responsibility, Scientists for 134
glossary of terms **10**, **47**
GM crops *see* genetically-modified crops
Gould, Steven J. 98
governance of science 47, 51
government advisors 105
government agencies, international 109
GRAINN technologies (Genomics, Robotics, Artifical Intelligence, Neuroscience, Nanotechnology) 10, 11, 15, 18-19, 47, 58, 67, 73, 72-6, 105
graphs 87, 88
Greece, classical 20-2
greenhouse effect 15
greenhouse gas emissions 84
Greenpeace 133

Index

gunpowder 36
halocarbons 93
Harvey, William 42
health see SHEE
Health, Environment and
    Justice, Center for 122, 133
HeLa 57
heresies 41
Hodgkin, Dorothy 5
Holocaust 98
homeopathy 132
Hooke, Robert 127
housing 120
human beings
    cell line contamination **57**
    cloning 32-3
    modification 131
Human Genome Project 53
humanitarian aims of sci-
    ence 97-9
human-ness 32
humility 37, 38, 45
Hydrogen cyanide 98
hydropower 68

iconic images 31
idealism 49
ignorance 75
    public 101, 102, 131
    scientific 89-93, 127
immune system 13
impersonality 62
impossibility, proof of 132
independence, scientific 48
India 22, 35, 56, 121, 128
induction 86
industry 11
    control of science 52
infallibility of science 38
information provision, public
    119-20
inquiry, disciplined 21
Institute for Science and
    Civilization 17
insulin 5
integrity 56, 58, 75, 105, 122
intellectual property 55
    see also patents
Intergovernmental Panel
    on Climate Change
    (IPCC) 84
Intermediate Technology
    Development Group 133
International Association for
    Science, Technology and
    Society 134
internet 122, 125, 128
    pathogens (malware) 14
inventions 103
inventors 96

investment see funding,
    research
IPCC see Intergovernmental
    Panel on Climate
    Change
Islamic civilization 22, 36
island countries
    sea levels **85**
ISO standards 57
isolation of scientists 49, 52
Italy 127

Japan 85
Jews
    genetic diseases **75**
    scientists 95
journalism, investigative
    59, 76, 93
Joy, Bill 11
Judaism 22
judgment
    exclusion of 63
    professional 72
    in statistical interpreta-
    tion 87-8
junk food
    addictions 16
    industry 121, 122

Kelvin, Lord 43, 44
Kepler, Johannes 24
Kiribati 85
Krimsky, Sheldon 104
Kuhn, Thomas S. 18, 65

labor, human
    replacement by robots
    12-13
Lack, Helen 57
land-use mapping 120
Lascaux cave paintings,
    France 34-5
Late lessons from early
    warnings 93
Lavoisier, Antoine
    Laurent 43
learning in mazes 59
lichens 44
light
    properties of 42
    velocity 29
    wave theory 43
lightning 81
little science 48-50
logging 32
logical positivist philoso-
    phy 95
Love Canal, US 122

M&M (malevolence and
    muddle) 10, 13-15, 47,
    90, 107
magic 27, 81, 82
Maldives 85
malevolence see M&M
malnourishment **121**
malware see internet:
    pathogens
Mamiraua project, Brazil 126
Manhattan Project 50
Marx, Karl 105
mathematics 21, 26, 38-9
May, Sir Robert 16
mechanics 26
media 104
    control 99
    independence 118
    research announce-
    ments 58
medical research 69, 102, 104
    racism 98
    see also pharmaceutical
    companies
medicine 26
    public influence 122-3
    see also complementary
    medicine
medicines see drugs,
    medicinal
mega science 52-60
    anti-democratic ten-
    dencies 106
'megaphone science' **59**
Mendel, Gregor 28
Mercury, orbits of 45
meta-analysis 104
Meteorological Office
    Hadley Centre, England 84
Methanol 98
Mexico 30
micro-finance 120
military control of science
    51, 53
The Mismeasure of Man 98
mobile telephones 108
    masts 67
Monsanto 112
moon, Galileo's **40**, 41
motives of science 11
muddle see M&M
multiculturalism 125
multinationals see corpo-
    rations
Murphy's Law 14, 15, 75, 92
music 21, 39

nanotechnology 10, 13, 108
    see also GRAINN
Naphthalene 98

Index

natural gas 68
natural selection 28, 43
Nazism 95, 98
need
   and addiction 109
Neem tree 56
negative results
   omission of 104
negotiation in good faith 119
Nelson-Rees, William 57
Netherlands 82, 119
New Zealand/Aotearoa 54
Newton, Isaac 24, 26, 42-3, 127
NGOs 113, 114-15, 118
non-scientists *see* public
nonviolence 125
'normal' science *see*
   applied science
Novak, J. 5, 6
nuclear bomb *see* atomic
   bomb
nuclear power 8
   civil 13, 52, 67, 68, 69,
   72, 83, 105
   military 83
nutrition science 122

obesity 16, **121**, 122
objectivity 61-77
On the Origin of Species 28
'open science' 76
openness in science 103-5
oppression
   and technologies 8
organic produce 122
outliers **64**, 65
overweight *see* obesity
ownership of research
   results 104, 118, 132
Oxford City Council
   Planning Committee,
   England 5
Oxford University,
   England 17
oxy-gene 43

paradigms 65
particle accelerators 52
Passover 22
Patent Co-operation Treaty
   applications **54**
patents 103
   plant **54**, 55, 106, 112
patients' campaigns 122-3
PCBs 93
peer review 48-9, 56, 57
permaculture 120
personal growth 125

personal questions 131-2
pesticides 31, 51
pharmaceutical compa-
   nies 104
   and biotechnology **106**
   see also medical
   research
philosophy 26, 43, 78
physics 11, 26, 52
plants
   patents **54**, 55, 106, 112
   see also biopiracy
plate tectonics 45
Plato 21
PNS *see* Post-Normal
   Science
poison gas 82
policy issues
   definition 100-1
policy-making
   public involvement in
   112-26
policy-related science
   99-103
politicians 118
politics
   local 5
   political questions 131
pollution 31, 67, 117, 122
   see also gender-bending
   pollutants
pollution-cleansing 13
poorer countries 125
   biopiracy 56
   technological develop-
   ment 109-11
Post-Normal Science (PNS)
   10, **18**, 19, 47, 69-73,
   91-3
Potter, Beatrix 44
poverty 97, 125
Precautionary Principle 10,
   89, 91-3, 103
predictions 13, 92
pressure groups 17
printing 36
prions 6
priorities, scientific 47, 131
privatization 53
probabilistic risk analysis 72
problems
   identification and solu-
   bility 5
   management 72-6
professional consultancy
   18, 71, 72
proof
   construction 87
   see also burden of proof
   property rights 53, 55
psychology 26

public
   awareness 69, 93
   ignorance 101, 102
   influence on technology
   107, **117**
   information provision
   99, 119-20
   and policy-making 112-26
   and scientific review 128-9
   trust 58, 78-9, 132
publication of research
   results 96, 104, 118, 128
Pugwash 50
pyramids of Egypt 35
Pythagoras 21, 38-9

Quakers 46
quality assurance 48, 56, 57
quantum theory 29
questions
   precautionary 89-90
   for students 129-32
   technology development
   108

racism 98
radiation 67
radioactive waste 70, **83**
radioactivity 44
rainbows 81
RAND 52
reductionism 65-6
Rees, Sir Martin 16
Reformation 22
relativity theory 30, 45
religion
   belief 22, 27, 28, 35
   domination of thought
   94-5
Renaissance 22
renewable energy research
   corporate funding **68**, 69
reproductive engineering 70
research 63
   agricultural 6
   and democracy 95-7
   temporary projects 59-60
research funding 93
   sponsorship 98
research results
   media announcements 58
   ownership 104, 118, 132
   publication of 96, 104,
   118, 128
research units 52
responsibility, social 50, 51
risk
   assessment 92
   management 100-1

Index

RIVM (Dutch Environmental Agency) 82, 133
road safety policy 101
robotics 10, 12-13
    see also GRAINN
Rose, Steven 59
Royal Commission on Environmental Pollution 6, 17
Royal Society of London 16
Russia 29, 95
Rutherford, Ernest 44

safety 103
    assurances of 80, 132
    safe limits 74
    standards 64
    see also SHEE
Saipan 85
Sardar, Zia 36
Scheele, Karl W. 43
schools
    pollution contamination 117
Schumacher, EF 120
science
    assumptions about 8, 130-1
    control of 14, 51, 52, 53, 128
    criticism of 26, 59, 65, 105, 118, 132
    and democracy 94-111
    Eastern science 35-6
    future of 11, 34, 127-32
    governance 47, 51
    historical errors 34-46
    humanitarian aims 97-9
    infallibility 38
    negative aspects 8
    nature of 127
    not undertaken 101
    objectivity 61-77
    priorities 47, 131
    social context 49-50
    threats to civilization 16
    Western science 20-33
Science and Development Network 134
Science in the Private Interest 104
science shops 119-20, 134
Scientific Method 48
Scientific Revolution 25-7
scientists
    anti-Nazi 95
    careers 52, 58, 96
    disagreement among 78-9, 132
    dogmatism 79
    egalitarianism 96
    as employees 104
    indoctrination 79
    Jewish 95
    recent existence of 126
    working in isolation 49, 52
Scientists, Union of Concerned 133
Scientists for Global Responsibility 60, 134
sea level rises 85
see-through science 76
senior citizens, isolation of 120
sexual orientation, respect for 125
Seychelles 85
SHEE (safety, health and environment, ethics) 10, 15, 16, 17, 47, 67, 89, 104, 108
Shell 68
sick school syndrome 117
Silent Spring 31, 32, 51, 82
smoke, tobacco
    chemical constituents 98
social context of science 49-50
social problems, solution of 97
Social Responsibility in Science campaigns 51
Socrates 91
software 14
    source code 128
solar power 13, 68
soy beans 112
space technology 95
species loss 70, 131
Spencer, Herbert 28
sponsorship of research 98
standard significance tests 102
Star Wars 53
Starry Messenger 24
state control of science 14
statistical surveys 86
statistical tests
    burden of proof 102-3
students
    questions for 129-32
Sun Microsystems 11
super-children 59
sustainability science 76
Switzerland 29
syphilis 98

Tay-Sachs disease 75
teaching 38, 45-6, 63, 132
    syllabus 65
    see also students
technology
    assessment 105
    dangers 8
    failures 13, 107
    illegal trade in 33
    illusions about 107
    and oppression 8
    public influence on 107-8
    unintended consequences 14
    see also nanotechnology
telescopes 24, 41
temperatures, global 84
terrorism, global 14, 84
textbooks 38, 49
theories, unchanging nature of 5
thermodynamics 29
Three Mile Island, USA 52, 72
Tibet 36
tides 41
time-measurement 22
tobacco companies 98, 99
tobacco smoke
    chemical constituents 98
Toluene 98
Tonga 85
trade in technology, illegal 33
travel, mass intercontinental 16
trends, 'best-fit' 64
Trinity College Cambridge, England 16
trust 58, 78-9, 132
truth 62-3, 97
Tudge, Colin 110
Tuvalu 85

uncertainty 5-6, 17, 18, 30, 31, 77, 78-93, 108
Union of Concerned Scientists 134
United States 31, 51, 52, 53, 54, 72, 83, 95, 96, 98, 99, 104, 106, 112, 115, 117, 120, 121, 122, 128
universe, Christian ideas of 22-3
unreliability 87
unsustainability 30
Urban VIII, Pope 42

value-neutrality 62, 65
vested interests 16, 92, 101, 104
Vietnam War 51
Vinyl chloride 98
Vulcanium 45

Index

Wallace, Alfred Russel 28
Waltner-Toews, David 126
waste 82
    radioactive 70, **83**
water 110
weapons 11, 51, 82
    see also atomic bomb
Wegener, Alfred 45
Western science
    origins of 20-33
whistle-blowers 93, 105

wind power 68, 120
WIPO *see* World Intellectual
    Property Organization
women
    in scientific careers 44, 96
World Intellectual Property
    Organization (WIPO) 54
World War One 81
World War Two 48, 50, 128
Worldwide Virtual Network
    of Young Practitioners

Working on Science and
    Society Issues 120, 133

yin-yang 43
yoga 35
Young Practitioners Working
    on Science and Society
    Issues, Worldwide Virtual
    Network of 120, 133
Yucca Mountain, US 83